# The Scientific Case for Atheism

## How Science, Suffering, and Reason Show That God Is Not Needed

## (Book 7)

---

**Dr Prabhat Das**

**MD, PhD, FACP**

# No God Required

## How Science Explains Reality Without the Divine

**Hardcover ISBN: 979-8-951067-18-0**

**Paperback ISBN: 979-8-951067-17-3**

**eBook ISBN: 979-8-951067-16-6**

Printed in the United States of America

# TABLE OF CONTENTS:

# Author's Note

## Why I Wrote This Series

Some books begin in libraries. Some begin in classrooms. This series began much earlier, in the mind of a curious boy growing up in India.

I was born into a Hindu religious family. Religion was not a distant theory in my childhood. It was part of daily life. It was present in prayers, rituals, festivals, stories, customs, family conversations, and the quiet assumptions that surrounded me. God was not discussed as a possibility. God was treated as reality.

My father was a firm believer in an omnipotent God. To him, God was real, powerful, protective, and deeply involved in human life. His belief was sincere. It was part of his strength, his discipline, and his view of the world. I grew up respecting that seriousness.

But from my teenage years, another force began to grow inside me.

Curiosity.

I wanted to know why things happened. I wanted to understand nature, disease, death, the stars, the human body, and the mind. I was not satisfied with an answer simply because it was old, sacred, or repeated by many people. I wanted to know whether it was true.

In high school, I became deeply impressed by the lives and writings of great scientists. I still remember going to the local library and borrowing books written by, or about, Newton, Darwin, Einstein, and other great thinkers. Those books were not popular among students at that time. In fact, I may have been the first student to borrow some of them. They were covered with dust. Some had even been damaged by insects.

But to me, they were treasures.

Those dusty books opened a new world.

Newton showed me that nature follows laws.

Darwin showed me that life has a history far deeper than ancient stories.

Einstein showed me that reality is stranger, deeper, and more beautiful than ordinary imagination.

These thinkers did not ask me to believe blindly. They invited me to understand. They did not frighten me with punishment. They opened the door to wonder. They showed me that truth does not become weaker when questioned. It becomes stronger.

That invitation changed my life.

My curiosity did not end with school. It continued through medical college, scientific training, clinical practice, and decades of work as a physician. In medicine, I learned the value of evidence. A diagnosis cannot be accepted because it feels comforting. A treatment cannot be trusted because it is traditional. A belief about the body must face reality. It must face observation, testing, correction, and outcome.

That same habit of mind gradually shaped the way I looked at religion and God.

After decades of medical practice and lifelong reading, I continue to be fascinated by science, philosophy, history, religion, psychology, evolution, cosmology, neuroscience, and the human mind. The more I read, the more clearly I see how far human knowledge has come. Many things once explained by God are now understood through natural causes. Thunder, disease, eclipses, famine, fertility, mental illness, death, life, consciousness, and the vast universe itself no longer need supernatural explanation.

This book, **The Scientific Case for Atheism: How Science, Suffering, and Reason Show That God Is Not Needed**, is the final volume in my seven-book series, **No God Required: How Science Explains Reality Without the Divine**.

The earlier books explored different parts of the same journey: how human beings created gods, how science replaced divine explanations, how the universe can be understood without a creator, how life evolved without design, how the mind does not require a soul, how morality does not need divine command, and how the personal God of religion fails when tested against suffering, prayer, miracles, and divine silence.

This final volume brings the case together.

It asks one central question:

Is God required?

My answer, after a lifetime of reading, thinking, practicing medicine, observing suffering, and studying science, is no.

God is not needed to explain the universe, life, the human body, disease, the mind, morality, religion, meaning, compassion, courage, wonder, or love.

I do not write these books to insult believers. Many believers are kind, sincere, moral, generous, and deeply thoughtful people. I know this from my own family, my culture, my patients, my friends, and my life experience.

Faith often gives people comfort, hope, discipline, community, and emotional strength.

But faith does not decide truth.

A belief may comfort us and still need examination. A tradition may be ancient and still be mistaken. A prayer may bring peace and still not prove that anyone is listening. A sacred story may be beautiful and still be human-made.

This final book is not an attack on people. It is an examination of claims.

If a personal God creates, designs, judges, reveals, heals, protects, answers prayer, and governs justice, then reality should show clear evidence of such a God. But when we look carefully, the pattern points in another direction.

The universe behaves naturally. Life behaves as evolution predicts. The body looks brilliant but imperfect. Disease follows biology, not sin. Suffering does not behave like justice. Prayer does not show reliable divine intervention. The mind changes when the brain changes. Morality grows from suffering, empathy, fairness, and reason. Religion follows geography, culture, childhood, power, fear, and hope.

Again and again, God becomes unnecessary.

This does not mean science has solved every mystery. It has not. We do not know everything about the origin of the universe, the origin of life, consciousness, or the deepest nature of reality. But an unanswered question is not evidence for God. Mystery should lead to inquiry, not premature certainty.

My goal in this series has been to bring together the insights of many great scientists, philosophers, historians, psychologists, physicians, and thinkers in language that an ordinary educated reader can understand. I hope these books encourage readers to think more rationally, question inherited beliefs, respect evidence, and study these subjects more deeply.

I also hope they show that life without God does not have to be empty, cold, selfish, or immoral.

A life guided by reason, compassion, truth, science, justice, and human responsibility can be deeply meaningful. We do not need divine command to care for one another. We do not need fear of hell to be moral. We do not need supernatural stories to feel wonder before the universe. We do not need to call suffering sacred in order to respond with love.

Human beings are already capable of kindness, courage, beauty, truth, and moral responsibility.

The universe is already astonishing.

Life is already precious.

Suffering is already urgent.

Our responsibility to one another is already clear.

No God is required for that.

# Introduction: The Final Question

## The Question This Book Must Answer

Every religion asks us, in one form or another, to trust that reality has a divine center.

The universe was created for a purpose. Life was designed. Human beings have souls. Suffering has meaning. Prayer reaches someone. Morality comes from above. Death is not the end. A hidden justice stands behind the visible world.

These ideas are powerful because they answer deep human needs.

They make the universe feel less indifferent. They make death feel less final. They make suffering feel less meaningless. They make morality feel protected by cosmic authority. They make human life feel watched, guided, and judged.

But there is one question that cannot be avoided:

Is any of this true?

That is the central question of this book.

This final volume of the **No God Required** series brings together the scientific, philosophical, moral, and human evidence examined throughout the earlier books. Its purpose is not to repeat every argument in full detail, but to ask what the whole pattern shows.

When we look honestly at reality, does the world look like it is governed by a personal God?

Or does it look like a natural world that human beings have tried to explain, comfort, and moralize through religion?

The answer matters because ideas about God are not harmless abstractions. They shape how people understand suffering, disease, morality, death, authority, guilt, meaning, and human responsibility.

If God is real, evidence should matter.

If God is not real, honesty should matter.

## The God Being Examined

The word God can mean many things.

For some people, God means nature. For others, God means love, mystery, energy, consciousness, beauty, or the universe itself. These poetic uses may express emotion, but they do not prove the God of religion. If God only means nature, then nature exists. If God only means love, then love exists. If God only means mystery, then the word God adds a name to what we do not yet understand.

This book examines a stronger claim.

It examines the personal God commonly defended by religion.

This is the God said to create the universe, design life, give human beings souls, reveal truth, command morality, answer prayer, perform miracles, judge human behavior, punish evil, reward faith, protect the innocent, and give meaning to suffering.

That God is not merely a symbol.

That God is a claim about reality.

Such a claim should leave evidence.

If a personal God governs the world, the world should not look exactly as it would look if no such God existed. We should expect some clear difference. We should expect more than inherited belief, emotional certainty, ancient texts, private experiences, and unexplained mysteries.

The question is not whether God is a comforting idea.

The question is whether God is a true explanation.

# Why the Standard Must Be Evidence

In every serious area of life, important claims require evidence.

A medical treatment must be tested. A diagnosis must fit the facts. A legal accusation must be supported. A scientific theory must explain observations. A safety warning must have reason behind it.

The God claim should not be exempt from this standard.

It is one of the largest claims ever made. It speaks about the origin of the universe, the purpose of life, the nature of the mind, the source of morality, the meaning of suffering, and the destiny of human beings after death.

A claim this large cannot rest only on faith.

Faith may comfort. Tradition may preserve. Scripture may inspire. Prayer may calm. Ritual may unite. Personal experience may feel powerful. But none of these automatically proves that a personal God exists.

Sincerity is not the same as truth.

Comfort is not the same as evidence.

Mystery is not the same as explanation.

This book does not ask readers to accept atheism blindly. It asks readers to examine the evidence with the same seriousness they would bring to any other important question.

If God exists, honest inquiry should not be dangerous.

If God does not exist, inherited belief should not be protected from examination.

## What the Whole Pattern Shows

The case for science-based atheism is cumulative.

No single chapter by itself carries the entire argument. The strength comes from the pattern that appears again and again when different areas of reality are examined.

The universe is vast, ancient, lawful, and mostly hostile to life. It does not place humanity at the center.

Life shows deep evidence of evolution: common ancestry, adaptation, extinction, imperfection, and biological struggle.

The human body is remarkable, but it is also fragile, flawed, and vulnerable to disease, cancer, genetic error, aging, and failure.

The origin of life remains a difficult scientific question, but an unanswered question is not evidence for God.

Suffering does not behave like justice. Children suffer without guilt. Disasters strike without moral selection. Animals suffered long before human beings existed.

Prayer may comfort, but it does not show reliable divine intervention.

The mind depends deeply on the brain. Memory, personality, language, judgment, emotion, and consciousness change when the brain changes.

Morality does not require divine command. It begins with the reality that conscious beings can suffer and flourish.

Religion itself follows human patterns: geography, childhood, culture, power, fear, hope, ritual, and identity.

This pattern does not point clearly toward a personal God.

It points toward a natural world.

## Why Suffering Cannot Be Treated Lightly

The question of God is not only about stars, cells, fossils, or philosophy.

It is also about suffering.

Any belief in a loving, all-powerful God must face the pain of the innocent. It must face children with cancer, newborns with genetic disease, families crushed by disaster, patients losing memory to dementia, and animals suffering across millions of years before human beings ever existed.

Religious explanations often try to give suffering a sacred meaning. It may be called a test, a punishment, a lesson, karma, divine plan, mystery, or soul-making.

But many of these explanations fail morally.

They may protect belief in God, but they can wound human beings. They may blame victims. They may make the sick feel guilty. They may ask grieving people to defend a divine plan they cannot understand.

A more honest response is possible.

Some suffering is not deserved.

Some suffering has no higher meaning.

Some suffering is simply the result of biology, nature, chance, cruelty, poverty, neglect, or disaster.

Our task is not to justify it.

Our task is to reduce it.

This is one of the moral strengths of science-based atheism. It does not need to defend God's reputation. It can focus fully on the suffering being.

## What Atheism Does Not Mean

Many people fear atheism because they misunderstand it.

Atheism does not mean hatred of believers.

It does not mean life is meaningless.

It does not mean morality is imaginary.

It does not mean beauty disappears.

It does not mean love becomes false.

It does not mean science has answered every question.

It does not mean human beings should become selfish or cold.

Science-based atheism is simpler than that. It is the conclusion that the God claim is not supported by sufficient evidence, and that reality is better explained through natural causes.

A person can reject belief in God and still live with compassion, gratitude, wonder, discipline, service, and moral seriousness.

The absence of God does not remove human responsibility.

It increases it.

If no divine power is guaranteed to rescue the suffering, then human care matters more. If no heaven is guaranteed to repair injustice, then justice matters here. If no sacred plan explains disease, then medicine, prevention, and research matter more. If no God is watching over humanity, then we must watch over one another.

This is not despair.

It is responsibility.

## How This Book Proceeds

This book begins with the foundation: what would count as evidence for God, and why faith cannot provide that evidence by itself.

It then examines the best arguments for God and asks whether they truly prove the God of religion. It looks at how science has steadily replaced supernatural explanations. It turns to the universe, life, the body, and the origin of life. It then moves to the deepest moral challenge: suffering, natural evil, divine hiddenness, prayer, and miracles.

From there, the book turns inward to the mind, asking whether the soul is needed. It then examines morality without God and religion as a human creation. Finally, it asks what human beings gain when they let go of unsupported belief and live honestly in the natural world.

The argument will not be that science has solved every mystery.

It has not.

The argument will be that God is not needed as an explanation.

We can admit uncertainty without turning it into theology. We can feel wonder without worship. We can build morality without divine command. We can face suffering without blaming the innocent. We can create meaning without heaven.

That is the case this book will make.

# Introduction Summary

1. This book examines the personal God of religion, not vague poetic uses of the word God.
2. A claim as large as God requires evidence, not only faith, tradition, comfort, scripture, or personal feeling.
3. The cumulative pattern of reality points toward natural explanation rather than divine action.
4. Innocent suffering is central because it challenges belief in a loving, all-powerful God.
5. Atheism does not remove meaning, morality, love, wonder, or responsibility.
6. This final volume brings together the scientific case that God is not needed to explain reality.

# Key Takeaway

If a personal God creates, designs, reveals, heals, protects, judges, and answers prayer, reality should show clear evidence of such a being. But the world behaves naturally. Life evolves. Bodies fail. Disease follows biology. Suffering is not morally distributed. Prayer lacks reliable evidence. Mind depends on brain. Morality grows from human life. Religion looks human-made. The evidence points toward one clear conclusion: God is not required to explain reality.

# Sources and Further Reading


1. Carl Sagan, **The Demon-Haunted World: Science as a Candle in the Dark**.
2. Bertrand Russell, **Why I Am Not a Christian**.
3. David Hume, **Dialogues Concerning Natural Religion**.
4. Charles Darwin, **On the Origin of Species**.
5. Richard Dawkins, **The Blind Watchmaker**.

6. Daniel C. Dennett, **Breaking the Spell: Religion as a Natural Phenomenon**.
7. J. L. Mackie, **The Miracle of Theism**.
8. J. L. Schellenberg, **Divine Hiddenness and Human Reason**.

## Transition to the Next Chapter

Before any claim about God can be accepted, we need a standard for judging it. What would actually count as evidence? What would distinguish a real personal God from tradition, emotion, coincidence, or mystery? The next chapter begins there, with the foundation of honest inquiry.

# Chapter 1: What Would Count as Evidence for God?

## Before Belief, We Need a Standard

Imagine that someone announces a remarkable medical cure.

He says this treatment can heal advanced cancer, reverse organ failure, prevent death, and work beyond anything modern medicine can explain. The claim is comforting. The stories are moving. Families share emotional testimonies. Some patients believe deeply that it helped them.

But as a physician, I would still have to ask:

Where is the evidence?

Did the treatment work better than no treatment?
Was the diagnosis correct?

Were other treatments used?
How many people improved?
How many did not?
Can the result be repeated?
Can independent observers verify it?
Are we looking at real cause, or only coincidence and hope?

No responsible doctor would accept such a large medical claim only because it felt meaningful, came from tradition, or made people feel better.

The same principle applies to God.

The claim that a personal God exists is one of the largest claims human beings can make. It is not a small private opinion. It is a claim about reality itself. It says there is an unseen conscious power behind the universe, life, mind, morality, prayer, suffering, death, and human destiny.

A claim this large needs a standard of evidence.

Before asking whether God exists, we must ask a simpler question:

What would count as good evidence?

## The Burden of Proof

In honest reasoning, the burden of proof belongs to the person making the claim.

If someone says an invisible being lives in his house, we do not have to disprove it before withholding belief. If someone says a certain stone cures disease, we do not have to prove that it never cured anyone anywhere. If someone says a spirit moves the planets, we are allowed to ask for evidence.

The same applies to God.

Atheism does not need to prove with absolute certainty that no possible God of any kind could exist. That is not how reasoning works. We do not believe every claim until it is disproved. We ask whether there is enough evidence to accept it.

The believer claims that God exists.

The believer must provide evidence.

This does not make atheism arrogant. It makes it cautious.

There is a great difference between saying, "I know everything," and saying, "I do not see enough evidence to believe this claim."

Science-based atheism is the second position.

It does not claim that every mystery is solved. It says that belief should be proportionate to evidence.

## Which God Are We Discussing?

The word God can be used in many ways.

Some people use it to mean the universe. Some use it to mean nature. Some use it to mean love, beauty, mystery, energy, consciousness, or moral feeling. These meanings may be poetic, but they do not prove the God defended by most religions.

If God means nature, then nature exists.
If God means love, then love exists.
If God means mystery, then mystery exists.

But renaming these things God does not establish divine action.

The question in this book is not whether human beings feel awe, love, wonder, or mystery. We clearly do.

The question is whether there is a real personal God who creates, designs, reveals, commands, hears prayer, performs miracles, judges human beings, and governs justice.

That kind of God should make a difference in the world.

If God is only a word for nature, then science already studies nature. But if God is an active conscious being, then we should ask what evidence shows that such a being exists.

Clarity matters because vague definitions can protect weak claims.

A God who answers prayer is very different from a God who simply means beauty.

A God who judges human beings is very different from a God who simply means the universe.

A God who reveals scripture is very different from a God who simply means mystery.

Before evidence can be judged, the claim must be clear.

## What Evidence Would Be Reasonable?

If a personal God existed and interacted with the world, several kinds of evidence could reasonably be expected.

First, there could be clear public evidence. God's existence would not depend mainly on birthplace, childhood teaching, or private feeling. People across cultures could recognize the same reality with reasonable clarity.

Second, prayer could show reliable patterns. If God answered prayer in the way many religions teach, careful study should reveal outcomes not explained by medicine, chance, psychology, or social support.

Third, miracles could be unmistakable. They would not have to hide in stories, rumors, unverifiable claims, or rare events that can be explained in several ways. A genuine miracle from an all-powerful God should be clear enough to distinguish from coincidence.

Fourth, revelation could be consistent. A God who wanted to guide humanity could provide moral and spiritual truth without leaving the world divided into thousands of conflicting religious claims.

Fifth, divine justice could be visible. A perfectly just God would not have to make the world look so much like a place where disease, disaster, and suffering strike without moral pattern.

Sixth, the mind might show independence from the brain. If an immortal soul were the true self, we might expect consciousness, memory, personality, and identity not to depend so completely on brain tissue.

These are not unreasonable demands.

They are the kinds of evidence we would expect if the claims were true.

A God powerful enough to create galaxies should not be limited to vague impressions, ancient writings, disputed miracles, and unexplained feelings.

## Evidence Should Be Public, Not Only Private

Private experience can feel powerful.

A person may feel God's presence during prayer. Another may feel guided during grief. Another may sense protection after surviving danger. Another may interpret a coincidence as a sign. These experiences may be emotionally real and important to the person who has them.

But private experience is not enough to prove a public claim about reality.

People of different religions report powerful experiences that support different gods, different scriptures, and different doctrines. A Hindu, Christian, Muslim, Sikh, Buddhist, or spiritual seeker may each feel deeply certain. Their experiences cannot all prove their conflicting beliefs.

The human mind is capable of powerful feeling, pattern-seeking, memory shaping, and interpretation. Under fear, grief, illness, loneliness, music, ritual, group emotion, or intense hope, experiences can feel larger than ordinary life.

That does not make the person dishonest.

It means personal experience must be interpreted carefully.

If God exists as a reality outside the mind, the evidence should not remain locked inside private feeling. It should be strong enough to examine beyond one person's interpretation.

## Evidence Should Distinguish God From Nature

Good evidence should separate one explanation from another.

If a patient recovers after prayer, we must ask whether the recovery also fits medicine, natural healing, misdiagnosis, immune response, timing, or chance.

If a person survives an accident, we must ask whether survival proves divine protection, or whether many others in similar accidents die despite prayer.

If a religious text contains wisdom, we must ask whether wisdom proves divine authorship, or whether human beings are capable of moral reflection.

If the universe inspires awe, we must ask whether awe proves God, or whether reality itself is astonishing without being supernatural.

This is essential.

An event is not evidence for God simply because it is emotional, unusual, beautiful, or not fully understood.

To count as strong evidence, it should point clearly to God rather than to natural causes, human psychology, chance, culture, or incomplete knowledge.

Weak evidence leaves many explanations open.

Strong evidence narrows the possibilities.

## Extraordinary Claims Need Strong Evidence

Some claims require ordinary evidence.

If someone says he owns a dog, we do not need a scientific study. The claim is common and fits ordinary experience.

But if someone says his dog can read minds, heal disease, or predict earthquakes, the standard rises. The claim is extraordinary. It goes far beyond ordinary experience. It requires stronger evidence.

God is an extraordinary claim.

Not because it is emotionally impossible, but because it proposes an unseen conscious being beyond nature who created or governs reality. It often includes claims about miracles, prayer, souls, revelation, judgment, heaven, hell, karma, or divine plans.

Such claims cannot be accepted with weak evidence.

A feeling is not enough. A tradition is not enough. A sacred book is not enough by itself. A coincidence is not enough. A gap in current knowledge is not enough.

The stronger the claim, the stronger the evidence must be.

This is not unfair to religion.

It is the same standard we use everywhere else when truth matters.

## What Does Not Count as Strong Evidence?

Several things may explain why people believe in God, but they do not prove that God exists.

A belief is not true simply because it is ancient. Many ancient beliefs were wrong.

A belief is not true simply because many people hold it. Large numbers of people have believed false things.

A belief is not true because it is comforting. Comfort can help us endure, but it does not decide reality.

A belief is not true because it is part of family or culture. Inherited beliefs can feel natural even when they are mistaken.

A belief is not true because a sacred text says so. A text must be examined before its authority is accepted.

A belief is not true because it fills a mystery. Mystery means we do not know. It does not mean we know God did it.

This point is simple but important.

The causes of belief are not the same as evidence for belief.

People may believe for emotional, cultural, historical, psychological, or family reasons. That is understandable. But truth requires more.

## The Problem With Special Pleading

Religious claims are often protected by a lower standard than other claims.

If a medical treatment fails, people question it.

If a business promise fails, people doubt it.

If a scientific prediction fails, the theory must be revised.

But if prayer fails, many believers say God had another plan. If suffering continues, they say the reason is mysterious. If miracles are not seen, they say faith is required. If religions conflict, they say only their own revelation is true. If evidence is weak, they say God cannot be tested.

This creates a protected belief.

The claim is made large enough to shape life, morality, and society, but then placed beyond ordinary examination.

That is not good reasoning.

If God is said to act in the world, then those actions can be discussed. If God answers prayer, prayer can be examined. If God heals, healing can be studied. If God reveals truth, revelation can be compared. If God governs justice, the world can be observed.

A claim cannot be both world-changing and evidence-proof.

If it makes claims about reality, it should face reality.

## Hiddenness as a Preview

There is another problem.

If God exists but leaves no clear evidence, the world begins to look very similar to a world without God.

Believers may say God is hidden for reasons we cannot understand. That answer will be examined more fully later. For now, we can simply note the difficulty.

A hidden God and no God can look almost the same.

Both leave us with unclear revelation.
Both leave us with unanswered prayer.
Both leave us with natural explanations.
Both leave us with suffering that follows no moral pattern.
Both leave us with religions shaped by culture and childhood.

If two explanations look the same in practice, we need a reason to choose one over the other.

Atheism chooses the simpler explanation: there is no convincing evidence for a personal God because such a God is not there.

This conclusion may be emotionally difficult, but it is intellectually honest.

## Why This First Question Matters

The question "What would count as evidence?" may seem simple, but it changes everything.

Without a standard, any belief can protect itself.

Every coincidence can become a sign. Every recovery can become a miracle. Every feeling can become God's presence. Every gap in knowledge can become divine action. Every unanswered prayer can be explained away. Every contradiction can be called mystery.

With a standard, belief must face reality.

This does not remove wonder. It protects truth.

It allows us to say: this is known, this is unknown, this is possible, this is unsupported, this is emotional, this is evidence, and this is only interpretation.

That kind of clarity is necessary before any serious discussion of God can begin.

The question is not whether belief feels meaningful.

The question is whether belief is justified.

# Chapter Summary

1. A claim as large as God requires a clear standard of evidence.
2. The burden of proof belongs to the person making the God claim.
3. The God being examined is not a vague symbol, but an active personal God who creates, reveals, judges, answers prayer, and governs reality.
4. Strong evidence should be public, testable, consistent, and able to distinguish God from nature, chance, psychology, and culture.
5. Ancient tradition, comfort, scripture, personal feeling, and mystery may explain belief, but they do not prove God.
6. A claim about divine action in the world should not be protected from examination.

# The Case So Far

The Introduction asked whether reality looks like it is governed by a personal God. This chapter established the first requirement for answering that question: we need a fair and serious standard of evidence before accepting so large a claim.

## Key Takeaway

Belief in God should not receive a lower standard of evidence than medicine, law, science, or ordinary life. If a personal God acts in reality, the evidence should be clearer than private feeling, inherited tradition, unexplained mystery, or ancient authority. Without strong evidence, disbelief is not arrogance. It is intellectual honesty.

## Sources and Further Reading


1. Carl Sagan, **The Demon-Haunted World: Science as a Candle in the Dark**.
2. David Hume, **An Enquiry Concerning Human Understanding**.
3. Bertrand Russell, **Why I Am Not a Christian**.
4. Antony Flew, **The Presumption of Atheism**.
5. Michael Shermer, **Why People Believe Weird Things**.
6. A. C. Grayling, **The God Argument**.


## Transition to the Next Chapter

Once a standard of evidence is established, the next question becomes unavoidable. Can faith meet that standard? The next chapter examines why faith can feel deeply convincing, yet still fail as evidence for truth.

# Chapter 2: Why Faith Is Not Evidence

## The Belief We Receive Before We Think

Most people do not first meet religion as an argument.

They meet it as childhood.

Before a child can examine evidence, he may already know the names of gods, prophets, scriptures, rituals, festivals, prayers, sins, blessings, and sacred stories. He learns what his family believes before he has the ability to compare beliefs. He learns what is holy before he knows how to ask whether it is true.

This is not the child's fault. It is how culture works.

Religion enters through love, authority, habit, repetition, music, fear, celebration, food, language, grief, and belonging. It becomes part of memory before it becomes a question.

That early power matters.

A belief taught with affection can feel like truth. A prayer repeated for years can feel like evidence. A story heard in childhood can feel deeper than reason. A religious identity shared by family and community can become emotionally difficult to question.

This is one reason faith feels so strong.

It is often planted before doubt is allowed to grow.

## What Faith Usually Means

The word faith can mean different things.

Sometimes people use faith to mean trust based on experience. A patient may have faith in a physician because the physician has training, skill, and a record of good care. A traveler may have faith that a bridge will hold because engineers designed it and many people cross it safely every day.

That kind of faith is not the problem.

It is trust supported by evidence.

Religious faith is often different. It asks people to believe in God, miracles, divine revelation, heaven, hell, karma, soul, or sacred authority even when strong public evidence is missing.

A person may be told:

Believe because scripture says so.
Believe because your ancestors believed.
Believe because doubt is dangerous.
Believe because life needs meaning.
Believe because God's ways are mysterious.
Believe because faith itself is a virtue.

This kind of faith does not test a claim.

It protects it.

That is why faith cannot serve as reliable evidence.

## Faith Can Support Opposite Beliefs

The greatest problem with faith is simple:

It can support many conflicting religions.

A Christian may have faith that Jesus is the Son of God.

A Muslim may have faith that Islam is the final revelation.

A Hindu may have faith in karma, rebirth, and many forms of the divine.

A Jew may have faith in the covenant and the God of Israel.

A Sikh may have faith in the teachings of the Gurus.

A Buddhist may have deep confidence in a path that does not require a creator God.

A person born into a traditional tribal religion may have faith in ancestral spirits.

Each believer may feel sincere. Each may have prayer, ritual, community, scripture, testimony, and emotional certainty. Each may say faith confirms the truth.

But these beliefs cannot all be true in the same way.

Faith can defend them all.

That means faith is not a reliable method for discovering truth.

A method that can lead different people to opposite conclusions cannot be trusted by itself.

## Sincerity Is Not Truth

Many believers are sincere.

Their devotion is real. Their prayers are heartfelt. Their sacrifices may be deep. Their moral discipline may be admirable. Their religious life may give them strength during suffering and courage during hardship.

This should be respected.

But sincerity does not prove truth.

A person can be sincerely wrong.

History is full of sincere beliefs that later proved false. People sincerely believed that the sun moved around the earth. They sincerely believed diseases were caused by spirits, curses, bad air, or divine punishment. They sincerely believed in omens, witchcraft, astrology, and sacred kings.

Their sincerity did not make those beliefs true.

The same standard must apply to religion.

A believer's sincerity tells us something about the believer.

It does not prove the belief.

## Comfort Is Not Proof

Religion often comforts people.

It may comfort the grieving by promising reunion after death. It may comfort the frightened by promising divine protection. It may comfort the guilty by offering forgiveness. It may comfort the suffering by saying pain has meaning. It may comfort the lonely by saying someone unseen is always present.

These comforts can be emotionally powerful.

But comfort does not prove truth.

A belief can reduce fear and still be false. A story can help someone endure loss and still be human-made. A prayer can calm the mind and still not prove that a divine being is listening.

In medicine, a comforting idea is not enough. A patient may feel better believing that an untested treatment will cure him. But a physician must still ask whether the treatment actually works.

The same applies here.

We can respect the emotional comfort of faith while still asking whether its claims are true.

Compassion does not require us to confuse comfort with evidence.

## Tradition Is Not Truth

Many religious beliefs survive because they are old.

People may say, "Our ancestors believed this," or "This has been taught for thousands of years," or "So many generations cannot be wrong."

But age does not prove truth.

Many ancient beliefs were mistaken. Ancient people explained thunder as divine anger, disease as curse or possession, eclipses as omens, infertility as spiritual punishment, and disasters as signs from heaven.

These explanations were old.

They were also wrong.

Tradition deserves historical respect, but it does not deserve automatic belief. Some traditions preserve wisdom. Some preserve beauty. Some preserve community. Some preserve error. Some preserve fear. Some preserve injustice.

Every tradition must be examined.

A belief does not become true because it has survived.

Sometimes it survives because it is supported by power, habit, fear, identity, and childhood teaching.

## Authority Is Not Evidence

Religion often depends on authority.

A priest says.
A scripture says.
A prophet says.
A guru says.
A tradition says.
A parent says.
A community says.

Authority can guide us when the authority is accountable to evidence. A physician, engineer, pilot, or scientist may deserve trust because their claims can be tested, corrected, and challenged.

Religious authority is often different.

It may claim protection from ordinary questioning. It may say doubt is sin, obedience is virtue, and questioning sacred teaching is pride. In such a system, authority does not guide inquiry. It stops inquiry.

This is dangerous.

No human being should be believed simply because he claims to speak for God. No book should be treated as beyond moral and factual examination simply because it is called sacred. No tradition should be immune from criticism because it is old.

Authority may explain why people accept a belief.

It does not prove the belief is true.

## Personal Experience Needs Interpretation

Many people say, "I have experienced God."

This deserves a careful response.

Personal experience can be meaningful. A person may feel peace during prayer, strength during grief, awe in nature, guilt after wrongdoing, or comfort in a crisis. Such experiences can shape a life.

But experience always needs interpretation.

A Christian may interpret peace in prayer as the presence of Jesus. A Hindu may interpret a similar peace as the grace of a deity. A Muslim may interpret it as nearness to Allah. A Buddhist may interpret it as the fruit of meditation. A secular person may interpret it as the mind calming through attention, ritual, breath, and emotional focus.

The feeling may be real.

The explanation may differ.

Human beings naturally search for patterns and meaning. We connect events. We remember coincidences. We notice what confirms our beliefs and forget what does not. In moments of grief, fear, danger, illness, or hope, the mind can create experiences that feel larger than ordinary life.

This does not make people dishonest.

It makes them human.

Private experience may be important to the person who has it, but it cannot settle a public question about reality.

## Faith Often Survives by Avoiding Testing

Strong beliefs should be open to examination.

But religious faith often protects itself from testing.

If prayer seems answered, it is taken as evidence for God.

If prayer is not answered, believers may say God had a different plan.

If good fortune comes, God is praised.

If tragedy comes, God's reasons are mysterious.

If scripture seems morally beautiful, it is called divine.

If scripture seems morally troubling, it is called symbolic, historical, or beyond human understanding.

If evidence appears weak, faith itself may be praised as the higher virtue.

This makes faith difficult to correct.

A belief that explains every outcome after the fact is not being tested. It is being protected.

In science, a claim must risk being wrong. If no possible observation could count against it, then it is not functioning as evidence-based knowledge.

Faith often avoids that risk.

That is why it can survive even when the world does not support it.

## Fear Can Protect Faith

Faith is often protected by fear.

A child may be told that doubt offends God. A teenager may be warned that questioning religion leads to moral collapse. An adult may fear hell, bad karma, family rejection, social shame, or loss of identity.

Fear makes belief harder to examine.

A person may not ask, “Is this true?” He may ask, “What will happen to me if I doubt?”

That is not free inquiry.

It is emotional pressure.

If a belief is true, it should not need fear to protect it. Truth can face questions. A real God would not be weakened by honest examination. A true explanation should become clearer under scrutiny, not require threats against the mind that questions it.

Fear may preserve religion.

It does not prove religion.

## Why Faith Feels Like Knowledge

Faith can feel like knowledge because it is supported by many emotional forces at once.

It may be repeated since childhood.

It may be shared by loved ones.

It may be tied to sacred music, rituals, festivals, and community.

It may comfort fear of death.

It may answer guilt.

It may give moral rules.

It may promise justice.

It may explain suffering.

It may provide identity.

When so many emotional needs gather around one belief, the belief can feel impossible to separate from life itself.

But emotional depth is not the same as factual truth.

A belief can feel deep because it is deeply rooted, not because it is true.

This is one of the hardest lessons in honest thinking.

The mind does not always experience inherited belief as inherited belief. It may experience it as obvious reality.

That is why faith must be examined carefully.

## The Difference Between Trust and Blind Belief

Some believers say, "Everyone has faith in something."

This is partly true, but it can be misleading.

We all rely on trust. We trust doctors, pilots, engineers, teachers, friends, maps, instruments, and memory. But reasonable trust is not blind belief.

Reasonable trust can explain itself.

A bridge is trusted because engineering principles work. A medicine is trusted because trials show benefit and harm. A pilot is trusted because training, instruments, regulation, and experience reduce risk. A scientific claim is trusted because evidence can be checked by others.

Religious faith often asks for belief without comparable evidence.

That difference matters.

The question is not whether human beings ever trust.

The question is whether the trust is proportional to evidence.

A trust that can be tested, corrected, and withdrawn is very different from a faith that must be preserved regardless of evidence.

## Science-Based Atheism Is Not Faith

Science-based atheism is often accused of being another faith.

But it is not faith in the religious sense.

It does not ask readers to believe without evidence. It does not offer sacred scripture. It does not threaten punishment for doubt. It does not require worship. It does not claim all mysteries are solved. It does not say scientists are infallible.

Science-based atheism says something more modest and more careful:

The God claim has not met its burden of proof.

It also says that natural explanations have repeatedly shown greater power than supernatural explanations.

Disease is better explained by biology than sin or demons.

Life is better explained by evolution than special creation.

Mind is better explained by brain than soul.

Religion is better explained by culture, psychology, and history than by divine revelation.

Suffering is better explained by nature, chance, and human action than by perfect divine justice.

This is not blind belief.

It is a conclusion drawn from evidence.

If strong evidence for God appeared, an honest atheist should examine it.

That openness is the difference between inquiry and faith.

## What Happens When Faith Is Tested

When faith is tested by evidence, it often changes shape.

Literal stories become symbolic.

Divine punishments become natural disasters.

Demonic possession becomes mental illness.

Sacred cosmology becomes metaphor.

Miracles become spiritual lessons.

Scriptural errors become matters of interpretation.

Failed prophecies become mysteries.

Unanswered prayers become hidden answers.

This movement may help believers preserve meaning, but it also reveals something important.

Faith often retreats when evidence advances.

The same belief that was once defended as literal may later be defended as symbolic. The same event once explained as supernatural may later be accepted as natural. The same scripture once treated as perfect may later need reinterpretation.

This pattern does not prove every religious claim false.

But it does show that faith often follows evidence only after resisting it.

A belief that must continually adjust to survive knowledge should not be treated as the foundation of knowledge.

## A More Honest Foundation

Human beings do not need to build life on unsupported faith.

We can build on evidence, reason, compassion, humility, and the courage to say "I do not know."

That may feel less comforting at first, but it is more honest.

It is honest to admit that we do not yet know everything about the universe.

It is honest to say that suffering is real and not always meaningful.

It is honest to accept that death is frightening without inventing certainty.

It is honest to question beliefs inherited in childhood.

It is honest to follow evidence even when it challenges comfort.

This kind of honesty does not remove wonder.

It deepens it.

A mind freed from compulsory faith can look at reality directly. It can ask questions without fear. It can admit uncertainty without shame. It can find meaning without pretending to know what it does not know.

That is a stronger foundation than faith.

## Chapter Summary

1. Most people receive religion in childhood before they can examine it critically.
2. Religious faith often means belief without strong public evidence.
3. Faith can support many conflicting religions, which makes it unreliable as a method for discovering truth.
4. Sincerity, comfort, tradition, authority, and personal experience may explain belief, but they do not prove it.
5. Faith often survives by avoiding testing and by using fear to discourage doubt.
6. Science-based atheism is not blind faith. It is the refusal to accept the God claim without sufficient evidence.

## The Case So Far

The previous chapter asked what would count as evidence for God. This chapter showed why faith cannot meet that standard. Faith may feel powerful because it is tied to childhood, identity, comfort, and fear, but it does not reliably distinguish truth from inherited belief.

## Key Takeaway

Faith can comfort, unite, and guide people emotionally, but it is not reliable evidence. Because faith can defend many conflicting religions, it cannot by itself show which belief is true. Honest inquiry requires more than sincerity, tradition, authority, or personal feeling. It requires evidence.

## Sources and Further Reading

1. Carl Sagan, **The Demon-Haunted World: Science as a Candle in the Dark**.
2. Michael Shermer, **The Believing Brain**.
3. Daniel C. Dennett, **Breaking the Spell: Religion as a Natural Phenomenon**.
4. Bertrand Russell, **Why I Am Not a Christian**.
5. A. C. Grayling, **The God Argument**.
6. Peter Boghossian, **A Manual for Creating Atheists**.

## Transition to the Next Chapter

If faith cannot prove God, the next step is to examine the strongest arguments offered in God's defense. Many of these arguments begin with real mysteries and serious questions. The next chapter asks whether they truly lead to God, or whether they make an unsupported leap from mystery to belief.

# Chapter 3: The Best Arguments for God and Why They Fail

## The Strongest Case Deserves a Fair Hearing

If we are going to reject belief in God, we should not reject a weak version of the argument.

We should examine the strongest version.

Religious belief is not held only by the ignorant or careless. Many intelligent, sincere, thoughtful people believe in God. Many have wrestled with deep questions about existence, morality, consciousness, beauty, suffering, and death. Their arguments deserve to be heard clearly before they are answered.

A serious atheist should not say, "All believers are foolish."

That is not true.

The better question is this:

Do the best arguments for God actually prove what believers think they prove?

Many arguments for God begin with something real. The universe exists. Life is complex. Moral experience is powerful. Consciousness is mysterious. Beauty moves us. Religious experiences feel profound. Human beings long for meaning and justice.

These are not trivial matters.

But a real question is not the same as a divine answer.

This chapter will examine the strongest arguments for God and show where they fail. Again and again, the problem is the same: the argument begins with mystery, order, beauty, or human longing, and then makes a leap to God.

That leap is not justified.

## The First Cause Argument

One of the oldest arguments for God begins with a simple question:

Why does anything exist at all?

Everything around us seems to have a cause. A child has parents. A tree grows from a seed. A house has builders. A planet forms from material under gravity. So, the argument says, the universe must also have a cause. That first cause is called God.

This argument has emotional force because existence itself is astonishing.

But it does not prove the personal God of religion.

First, if everything needs a cause, then God would also need a cause. If God does not need a cause, then the rule "everything needs a cause" has already been abandoned. The believer has simply made an exception for God.

Second, even if the universe has some deeper cause, we do not know that the cause is a conscious being. It could be a physical reality we do not yet understand. It could be a feature of cosmology beyond present knowledge. It could be something for which our ordinary idea of cause is inadequate.

Third, even if there were a first cause, that would not prove prayer, miracles, scripture, soul, heaven, hell, divine judgment, or a loving God involved in human life.

At most, the first cause argument points to a mystery about existence.

It does not give us the God of religion.

## The Fine-Tuning Argument

The fine-tuning argument says the universe appears to have physical constants that allow life to exist. If certain values were different, stars, planets, chemistry, or life might not have formed. Therefore, the argument says, the universe must have been designed by God.

This argument deserves attention.

The conditions that allow life are indeed remarkable. The universe is intelligible. Physical laws are deep and elegant. Life depends on a delicate structure of matter, energy, chemistry, and time.

But fine-tuning does not prove God.

We do not know whether the constants of nature could have been different. We do not know whether deeper physical laws explain them. We do not know whether many universes exist with different properties. We also must remember that observers can exist only in a universe compatible with observers. We should not be surprised to find ourselves in a universe where our existence is possible.

There is another problem.

If the universe was designed for life, it is an extremely strange design.

Most of the universe is empty, cold, airless, radioactive, and deadly. Life survives on a tiny planet around one ordinary star in a vast cosmos mostly hostile to biology. Even on Earth, life survives through struggle, predation, extinction, disease, and accident.

The universe may permit life.

It does not look lovingly designed for life.

Fine-tuning raises questions. It does not prove a personal God.

## The Design Argument From Life

Before Darwin, living things seemed to provide powerful evidence for design.

The eye sees. The wing flies. The heart pumps. The seed becomes a tree. The body heals. Life appears organized, purposeful, and complex.

It is easy to understand why people saw God in living things.

But evolution changed the question.

Natural selection explains how complex adaptations can arise without a designer. Variation occurs. Some traits help organisms survive and reproduce. Those traits become more common across generations. Over long periods, small changes can build structures that appear designed.

Life looks designed because natural selection produces the appearance of design.

Evolution also explains what perfect design struggles to explain: poor anatomy, genetic disease, vestigial structures, extinction, parasites, cruelty in nature, and biological waste.

A perfect designer is not needed to explain life's complexity.

A long natural process explains it better.

The design argument from life once seemed strong because people lacked a natural mechanism. Darwin provided that mechanism. Since then, genetics, fossils, comparative anatomy, embryology, and DNA have strengthened the evolutionary explanation.

Life is beautiful.

But beauty is not proof of design.

## The Moral Argument

Another common argument says morality requires God.

If there is no God, people ask, what makes anything right or wrong? Why should cruelty be condemned? Why should honesty, compassion, or justice matter?

This concern is understandable.

Morality feels deep. When we see cruelty, we do not feel that it is merely a personal preference. We feel that something real is wrong.

But God does not solve the problem as easily as believers often think.

If something is good only because God commands it, then morality becomes arbitrary. If God commanded cruelty, would cruelty become good? Most believers would say no.

But if God commands what is good because it is already good, then goodness does not depend entirely on God's command.

Morality is better understood through the reality of conscious life.

Beings can suffer. Beings can flourish. Pain, fear, hunger, love, trust, humiliation, kindness, and cruelty are real experiences. Moral concern begins because our actions affect lives that can be harmed or helped.

God is not needed to make suffering matter.

The moral argument points to a real human concern, but it does not prove divine command.

## The Consciousness Argument

Consciousness is one of the deepest mysteries.

How does matter produce experience? How does the brain create awareness, memory, selfhood, intention, emotion, and imagination? Why is there something it feels like to be alive?

Because consciousness is difficult, some people argue that it must come from God, soul, or cosmic mind.

But difficulty is not evidence.

The brain is strongly connected to the mind. When the brain changes, consciousness changes. Stroke can remove speech. Dementia can erase memory. Brain injury can alter personality. Anesthesia can turn awareness off. Drugs can change mood, thought, and perception.

These facts do not solve every mystery of consciousness, but they point strongly toward a natural dependence of mind on brain.

If the soul were the true independent self, it is difficult to explain why the self changes so deeply when the brain is damaged.

The consciousness argument begins with a real mystery.

But a mystery about mind does not prove God.

It tells us that neuroscience and philosophy still have work to do.

## The Argument From Religious Experience

Many people believe because they feel they have experienced God.

They may feel divine presence in prayer, peace during grief, awe in nature, forgiveness after guilt, or guidance during crisis. Such experiences can be powerful and life-changing.

They should not be mocked.

But they must be interpreted carefully.

People from many religions report experiences that confirm different beliefs. A Christian may feel Jesus. A Muslim may feel Allah. A Hindu may feel the presence of a deity. A Buddhist may feel deep awakening without a creator God. A spiritual person may feel universal energy. A secular person may feel awe without interpreting it as supernatural.

The feeling may be real.

The explanation may be wrong.

Human beings are meaning-making creatures. We seek patterns, remember coincidences, respond to ritual, and interpret emotion through beliefs we already have. Religious experience often follows religious expectation.

If experiences from different religions support conflicting conclusions, experience alone cannot prove which religion is true.

Religious experience may show the depth of the human mind.

It does not prove God.

## The Argument From Scripture

Many believers say God is known through scripture.

The Bible, Quran, Vedas, Upanishads, Guru Granth Sahib, Buddhist texts, and many other sacred writings have shaped civilizations and changed lives. They contain poetry, law, myth, moral teaching, ritual, history, devotion, and spiritual reflection.

Some passages are beautiful and wise.

But scripture cannot simply prove itself.

Every religion can appeal to its own sacred text. A book claiming divine authority does not become divine merely because it says so or because a community treats it as holy.

Scriptures also bear human fingerprints. They are written in human languages, shaped by human cultures, preserved by human institutions, interpreted by human authorities, and marked by the moral limits of their time.

If a scripture contains wisdom, that shows human beings can produce wisdom.

If it contains troubling moral passages, believers must use human judgment to reinterpret or reject them.

Scripture may be important as literature, culture, devotion, or history.

It is not enough by itself to prove God.

## The Argument From Beauty and Wonder

Some people feel that beauty points to God.

A night sky, a child's face, a mountain, a forest, a bird in flight, music, mathematics, or an act of kindness can produce awe so deep that it seems to demand a divine source.

This feeling is understandable.

Beauty is one of the most powerful parts of human life.

But beauty is not proof of God.

The world contains beauty and terror. A sunset may be beautiful, but the same natural world contains cancer, parasites, earthquakes, predation, starvation, and decay. If beauty points to a loving creator, what does pointless suffering point to?

Beauty may reveal the sensitivity of the human mind. It may reflect patterns in nature, the structure of perception, the emotional power of music, or the deep relationship between living beings and the world around them.

Understanding beauty naturally does not destroy it.

A rainbow is not less beautiful because we understand light.

A flower is not less beautiful because we understand evolution.

The stars are not less beautiful because we understand physics.

Wonder does not require worship.

## The Argument From Human Longing

Human beings long for meaning, justice, love, forgiveness, immortality, and reunion with those who have died.

Religion answers these longings.

It says death is not final. Justice will come. Love can last forever. Suffering has meaning. The universe is not indifferent. Someone sees us. Someone cares.

These promises are emotionally powerful.

But longing does not prove reality.

Hunger points to food because food exists and bodies evolved to need it. But not every longing has an object that exists. Human beings may long for immortality because we fear death. We may long for perfect justice because injustice wounds us. We may long for cosmic love because life can feel lonely and fragile.

A wish can be understandable and still remain a wish.

The fact that human beings desire God does not prove God.

It proves that human beings desire meaning, comfort, justice, and survival.

Religion may answer longing.

That does not make the answer true.

## The Unjustified Leap

The main weakness in these arguments is not that they ask foolish questions.

They often ask serious questions.

Why is there something rather than nothing?
Why does the universe permit life?
Why does life appear designed?
Why do we experience moral obligation?
Why is consciousness so mysterious?

Why do religious experiences feel powerful?
Why does beauty move us?
Why do we long for meaning beyond death?

These are real questions.

But the argument for God usually makes a leap.

From existence to creator.
From order to designer.
From beauty to divine artist.
From morality to lawgiver.
From consciousness to soul.
From longing to heaven.
From mystery to God.

The leap is too large.

Even if we admitted that some deep mystery remains, it would not follow that a personal God exists. It would not follow that this God answers prayer, reveals scripture, performs miracles, judges souls, creates heaven or hell, or cares about human rituals and beliefs.

A question mark is not a throne.

Mystery does not become God simply because we give it a sacred name.

## Why These Arguments Continue to Persuade

If these arguments fail, why do they continue to persuade so many people?

Because they connect with deep human needs.

The first cause argument speaks to our need for explanation.

Fine-tuning speaks to our need for cosmic meaning.

Design speaks to our sense that life is too beautiful to be accidental.

Morality speaks to our need for justice.

Consciousness speaks to our sense that the self is more than matter.

Religious experience speaks to our need for presence and comfort.

Scripture speaks to our need for authority.

Beauty speaks to our need for wonder.

Longing speaks to our fear of death and desire for love to last.

These needs are real.

But emotional power is not the same as evidence.

The best arguments for God often feel strong because they begin where human beings are most vulnerable: mystery, beauty, fear, morality, suffering, and death.

That is why they deserve compassion.

But they still need examination.

## A More Honest Answer

It is honest to say that the universe is mysterious.

It is honest to say that consciousness is difficult.

It is honest to say that beauty moves us.

It is honest to say that morality matters deeply.

It is honest to say that death frightens us.

It is honest to say that human beings long for justice, meaning, and love.

But honesty does not allow us to turn every deep question into God.

A mature mind can live with unanswered questions. It can keep studying the universe without inventing a creator. It can value morality without divine command. It can feel wonder without worship. It can face death without pretending to know what lies beyond it.

The failure of these arguments does not make reality empty.

It makes reality open to honest inquiry.

The strongest case for God begins with real questions, but it does not reach the conclusion religion needs.

The God of religion remains unproven.

## Chapter Summary

1. The strongest arguments for God begin with real questions about existence, order, life, morality, consciousness, beauty, experience, and longing.
2. The first cause argument does not prove a personal God.
3. Fine-tuning raises questions but does not establish divine design or religious truth.
4. Evolution explains the appearance of design in life better than special creation.
5. Morality, consciousness, religious experience, scripture, beauty, and longing do not require God.
6. The central weakness in these arguments is the unjustified leap from mystery or meaning to the God of religion.

## The Case So Far

Chapter 1 established the need for evidence. Chapter 2 showed why faith cannot provide that evidence. This chapter examined the strongest arguments offered in God's defense and found that they raise serious questions, but do not prove the personal God claimed by religion.

## Key Takeaway

The best arguments for God often begin with real mystery, beauty, morality, consciousness, or longing. But they fail when they leap from those human experiences to a personal God who creates, reveals, judges, answers prayer, and governs reality. A deep question is not the same as divine evidence.

## Sources and Further Reading

1. David Hume, **Dialogues Concerning Natural Religion**.
2. Bertrand Russell, **Why I Am Not a Christian**.
3. J. L. Mackie, **The Miracle of Theism**.
4. Richard Dawkins, **The Blind Watchmaker**.
5. Daniel C. Dennett, **Consciousness Explained**.
6. A. C. Grayling, **The God Argument**.
7. Graham Oppy, **Arguing About Gods**.

## Transition to the Next Chapter

The arguments for God often grow stronger when people do not yet have natural explanations. But history shows a clear pattern: where gods once explained the world, science gradually uncovered nature. The next chapter follows that retreat and examines the shrinking space for God.

# Chapter 4: The Shrinking Space for God

## When Fear Became Explanation

Imagine a village centuries ago.

A child suddenly falls to the ground, stiffens, shakes violently, and becomes unconscious. His family is terrified. No one knows about neurons, electrical activity in the brain, epilepsy, fever, infection, low blood sugar, or head injury. No one has an EEG machine. No one has modern medicine.

What explanation is available?

A spirit has entered him.
A demon has attacked him.
God is angry.
A curse has been placed on the family.
A ritual must be performed.

To the people of that time, such explanations did not feel foolish. They felt natural. When human beings faced events they could not understand, they filled the unknown with invisible agency. Someone unseen must be acting. Someone must be responsible. Some hidden power must be behind the frightening event.

This is one of the oldest habits of the human mind.

When thunder roared, the gods were angry.
When crops failed, the gods withheld favor.
When disease spread, sin or spirits were blamed.
When eclipses darkened the sky, heaven was sending a sign.
When a child died, people searched for a sacred reason.

Before science, supernatural explanation was not rare. It was normal.

Gods, spirits, demons, curses, omens, and miracles lived in the spaces where knowledge was missing.

But as human knowledge grew, those spaces began to shrink.

## The Ancient World Was Full of Divine Causes

Ancient human beings lived close to danger.

Storms could destroy homes. Drought could bring famine. Infection could kill entire families. Childbirth could take the life of mother or child. Animals could attack. Crops could fail. A small wound could become fatal. A fever could appear suddenly and disappear mysteriously, or end in death.

Life was uncertain, and uncertainty demands explanation.

Without science, people often explained the world through intention. If something happened, someone must have wanted it. If rain came, a god was pleased. If rain did not come, a god was angry. If disease struck, someone had sinned, a spirit had acted, or fate had turned.

This way of thinking was deeply human.

Human beings are social animals. We understand other people by reading intention: Who is angry? Who is helping? Who is threatening? Who is responsible? That skill helped our ancestors survive. But it also made them see intention where no intention existed.

Nature was treated like a person.

The sky could be angry. The river could be sacred. The mountain could demand sacrifice. The disease could be punishment. The earthquake could be warning. The unusual birth could be omen.

Religion grew in this world of fear, uncertainty, and imagined agency.

It gave explanations before evidence was available.

## Lightning, Thunder, and the Sky

The sky was one of the first great homes of the gods.

Thunder sounded like power. Lightning looked like a weapon. Eclipses seemed terrifying. Comets appeared without warning. The sun gave life. The moon shaped time. Stars guided travelers and farmers. The sky felt far above ordinary human life, so it easily became the place of divine activity.

Many cultures imagined gods of thunder, rain, sun, moon, storm, and fertility.

These beliefs made emotional sense. A storm can still make a modern person feel small. To ancient people, without meteorology, electricity, astronomy, or satellites, the sky must have seemed alive with intention.

But science changed that.

Lightning became an electrical discharge. Thunder became the sound of rapidly expanding heated air. Eclipses became predictable alignments of sun, moon, and Earth. Comets became objects moving through space. Seasons became the result of Earth's tilt and orbit. The stars became distant suns, not lamps placed in a dome for human guidance.

The sky did not become less beautiful.

It became better understood.

The gods did not explain lightning once electricity was known. Divine anger was no longer needed to explain thunder. Omens were no longer needed to explain eclipses.

A natural explanation replaced a supernatural one.

# Disease Before Medicine

Disease was another great territory of the supernatural.

Before germ theory, genetics, physiology, pathology, and public health, illness often looked mysterious and personal. Why did one child die while another lived? Why did fever spread through a village? Why did a woman die in childbirth? Why did a wound turn black? Why did the mind become confused? Why did plague strike one city and not another?

Religious answers came easily.

Sin. Curse. Possession. Divine punishment. Karma. Impurity. Witchcraft. Evil eye. Bad spirits. Broken taboo.

These explanations gave people a sense of order, but often at a terrible moral cost. The sick were not only sick. They could be blamed. The poor were not only poor. They could be seen as spiritually deserving their condition. The mentally ill were not only suffering. They could be treated as possessed or cursed.

Medicine gradually changed this picture.

Infection became linked to microorganisms. Epidemics became connected to sanitation, transmission, crowding, water, insects, and public health. Diabetes became a disorder of insulin and metabolism. Tuberculosis became a bacterial disease. Seizures became disturbances of brain activity. Cancer became uncontrolled cell growth. Genetic diseases became linked to DNA.

The more medicine learned, the less disease needed supernatural explanation.

This did not remove suffering, but it changed the moral response.

A patient with infection needs treatment, not blame.

A child with epilepsy needs care, not exorcism.

A person with depression needs compassion and help, not accusation of spiritual weakness.

Natural explanation made medicine more humane.

## Mental Illness and Possession

Few areas show the moral importance of natural explanation more clearly than mental illness.

For much of human history, unusual behavior was often explained through spirits, demons, curses, moral failure, or divine punishment. A person hearing voices, behaving strangely, becoming severely depressed, losing touch with reality, or having seizures could be feared as possessed or polluted.

Such beliefs did not merely explain suffering.

They often increased it.

People with mental illness could be isolated, chained, beaten, shamed, exorcised, or treated as spiritually dangerous. Families could hide them. Communities could fear them. Patients could be blamed for symptoms they did not choose.

Modern knowledge has not solved every problem of mental illness. Much remains difficult. But it has changed the foundation. We now understand that brain chemistry, trauma, genetics, development, stress, sleep, substances, medical illness, and social conditions can all affect the mind.

This natural understanding does not make patients less human.

It makes them more deserving of care.

A brain-based explanation can reduce shame. It can lead to treatment, therapy, medication, social support, and compassion.

When supernatural explanations retreat, human responsibility often grows.

## The Body Became Natural

The human body once seemed full of divine mysteries.

Why did the heart beat?
Why did wounds heal?
Why did women bleed monthly?
Why did pregnancy happen?
Why did pain travel?
Why did fever rise?
Why did old age weaken the body?
Why did death come?

For centuries, answers were mixed with spirits, humors, sacred forces, sin, fate, or divine will.

Modern biology transformed this understanding.

The heart became a pump. The lungs exchanged gases. The kidneys filtered blood. Hormones carried chemical messages. Nerves transmitted signals. The immune system defended and sometimes misfired. Cells divided. Genes carried information. Mutations occurred. Organs failed for identifiable reasons.

The body did not become less amazing.

It became more amazing.

To understand the heart as muscle, electricity, valves, pressure, rhythm, oxygen, and circulation is not to make it dull. It is to see its real wonder. To understand the immune system as cells, antibodies, inflammation, memory, and regulation is not to remove mystery. It is to replace vague mystery with deeper knowledge.

Science did not steal the body from God.

It rescued the body from ignorance.

And in doing so, it saved lives.

## Life Before Darwin

Before Darwin, the living world seemed to be the strongest evidence for design.

The bird's wing, the eye, the flower, the seed, the spider's web, the human hand, the beating heart, and the nursing infant all appeared purposeful. Their complexity seemed too great to arise naturally.

So people concluded that a designer must have made them.

This was understandable.

Without evolution, the design argument felt powerful. Living things really do appear organized. They fit their environments. They reproduce. They adapt. They seem built for function.

Darwin changed the question.

He showed how natural selection could produce the appearance of design without a designer. Living things vary. Some variations help survival and reproduction. Those traits become more common over generations. Given enough time, simple changes can produce complex adaptation.

Later, genetics strengthened this explanation. DNA showed how traits are inherited and modified. Fossils showed change across time. Comparative anatomy showed shared structures. Embryology showed developmental connections. Modern biology showed that life is related through common ancestry.

Life did not need separate acts of creation.

It had a natural history.

This was one of the greatest reductions in the explanatory role of God.

## Imperfection Became Evidence

The retreat of design became even clearer when scientists began to understand biological imperfection.

A perfect designer might be expected to create clean, efficient, painless, and morally balanced life. But actual life is full of compromise.

The human spine carries the cost of upright walking.
The throat allows choking because breathing and swallowing share a dangerous crossing.
The eye has a blind spot.
The appendix can become inflamed.
Wisdom teeth often do not fit.
Genes mutate.
Cells become cancer.
The immune system can attack the body.
Species go extinct.
Predators tear prey apart.
Parasites live by harming other organisms.

These facts fit evolution better than perfect design.

Evolution does not produce perfection. It works with what already exists. It modifies old structures. It trades one advantage against another. It preserves what helps reproduction, not what prevents all suffering.

Once seen this way, imperfection is no longer surprising.

It is expected.

The body is not a machine designed from scratch by a perfect engineer. It is a historical organism shaped by ancestry, selection, accident, and compromise.

That insight changed the meaning of life's flaws.

They became evidence for evolution, not problems to hide from design.

## The God of the Gaps

As science explained more, a pattern appeared.

God was placed in the gaps.

When lightning was unexplained, God or gods explained it.
When disease was unexplained, spirits or punishment explained it.
When life's complexity was unexplained, design explained it.
When consciousness remained mysterious, soul explained it.
When the universe seemed mysterious, creation explained it.

But each time natural knowledge grew, the supernatural explanation retreated.

This is called the God of the gaps.

The problem is not merely that the gaps shrink. The deeper problem is that a gap in knowledge is not evidence for God. It only means we do not yet know. To place God in the gap is to build belief on ignorance.

That is unsafe ground.

History shows that many gaps once filled with God later became fields of science.

A wise person should be cautious before placing God in the next gap.

## “Science Cannot Explain Everything”

Believers often reply, “Science cannot explain everything.”

That is true.

Science has not answered every question. We do not fully understand the origin of the universe, the origin of life, consciousness, dark matter, dark energy, or the deepest nature of reality.

But incomplete science does not prove God.

To say “science cannot explain this yet” is honest. To say “therefore God” is not.

The word “yet” matters.

Many things that were once unexplained became explainable. No one in the ancient world could have explained bacteria, DNA, plate tectonics, electricity, anesthesia, oxygen, black holes, or the age of Earth. Their ignorance did not make supernatural explanations correct.

Science is a method, not a complete book of answers.

Its strength is that it keeps asking, testing, revising, and learning.

Religion often stops too soon.

It gives an answer before the evidence is complete and then protects that answer from correction.

## Science Changes the Moral Response

The shrinking space for God is not only an intellectual matter.

It is also moral.

When suffering is explained supernaturally, people may blame the victim. A disease may be called punishment. Infertility may be called divine displeasure. Mental illness may be called possession. Poverty may be called karma. Disaster may be called judgment.

Such explanations can deepen suffering.

Natural explanation changes the response.

If cholera spreads through contaminated water, we clean the water.
If malaria spreads through mosquitoes, we fight mosquitoes.
If cancer grows from cell mutations, we study cells.
If epilepsy comes from brain activity, we treat the brain.
If depression involves brain, body, trauma, and life stress,

we offer care.
If disasters happen through geology and weather, we prepare, warn, rescue, and rebuild.

Science does not solve every human problem, but it points us toward action.

It replaces blame with investigation.

It replaces ritual fear with prevention.

It replaces divine punishment with human responsibility.

That is one of the great moral achievements of natural explanation.

## Science Does Not Remove Wonder

Some fear that science makes the world cold.

But this fear misunderstands science.

Understanding lightning does not make the storm less majestic. Understanding the moon does not make it less beautiful. Understanding the heart does not make life less precious. Understanding evolution does not make animals less astonishing. Understanding the brain does not make love less real.

Science removes false explanations.

It does not remove wonder.

In many cases, it deepens wonder.

A rainbow becomes more beautiful when we understand light, water, and perception. A tree becomes more astonishing when we understand photosynthesis, roots, fungi, cells, and evolution. The night sky becomes more moving when we understand that stars are suns, galaxies are vast, and the atoms in our bodies came from ancient cosmic processes.

The supernatural often gives quick answers.

Science gives deeper understanding.

Wonder does not need ignorance.

## The Shrinking Space for God

Across history, the pattern is clear.

The sky became astronomy and meteorology.

Disease became biology and medicine.

Mental illness became neuroscience, psychology, and social care.

The body became anatomy, physiology, genetics, and pathology.

Life became evolution.

Disasters became geology, climate, weather, and risk science.

Again and again, divine explanation lost territory.

This does not mean science has answered every question. It has not. But the direction of discovery has been unmistakable. Natural explanations have expanded. Supernatural explanations have retreated.

God has not become clearer as knowledge has grown.

God has become less necessary.

This is not because scientists set out to destroy God. It is because nature, when studied carefully, keeps explaining itself through natural causes.

That is the historical pattern.

And it prepares us for the next question.

If God is no longer needed to explain lightning, disease, mental illness, the body, or life's diversity, is God still needed to explain the universe itself?

# Chapter Summary

1. Early human beings often explained frightening or mysterious events through gods, spirits, curses, demons, or divine punishment.
2. Science gradually replaced supernatural explanations in the sky, disease, mental illness, the body, and life.
3. Darwin's theory of evolution greatly reduced the need for divine design in biology.
4. Biological imperfection fits evolution better than perfect design.

5. The God of the gaps is weak reasoning because ignorance is not evidence.
6. Natural explanation often improves morality by replacing blame with care, prevention, and responsibility.
7. Science does not remove wonder. It deepens wonder by replacing false explanations with understanding.

## The Case So Far

The earlier chapters established that God claims require evidence, faith is not evidence, and the best arguments for God make an unjustified leap. This chapter showed that history has repeatedly moved explanation from the supernatural to the natural.

## Key Takeaway

God once occupied many spaces in human ignorance: storms, disease, mental illness, the body, life, disaster, and death. As knowledge grew, those spaces shrank. Science has not explained everything, but history shows a clear pattern. Natural explanations keep expanding, while supernatural explanations retreat.

## Sources and Further Reading

1. Carl Sagan, **The Demon-Haunted World: Science as a Candle in the Dark**.
2. Steven Weinberg, **To Explain the World**.
3. David Wootton, **The Invention of Science**.
4. Charles Darwin, **On the Origin of Species**.
5. Richard Dawkins, **The Blind Watchmaker**.

6. Edward Shorter, **A History of Psychiatry**.
7. Roy Porter, **The Greatest Benefit to Mankind**.

## Transition to the Next Chapter

For many believers, the last great refuge of God is not thunder, disease, or life's diversity, but the universe itself. If science has explained so much within the world, can it also make sense of the world as a whole? The next chapter turns from Earth to the cosmos.

# Chapter 5: The Universe Without a Divine Hand

## The Sky That Once Belonged to God

For most of human history, the sky felt close to God.

The heavens were above. Human beings were below. The sun gave warmth. The moon measured time. Stars guided travelers. Storms frightened villages. Eclipses darkened the day. Comets appeared like warnings. The sky seemed vast, mysterious, and powerful.

It was natural for ancient people to place gods there.

The heavens felt like a realm of intention. Something above us seemed to watch, rule, bless, punish, or speak. The word heaven itself came to mean both sky and divine home.

Modern science changed that picture.

The sky did not disappear. Its beauty did not fade. But its meaning changed. The sun became a star. Earth became a planet. The moon became a natural satellite. The stars became distant suns. The Milky Way became one galaxy among many. Space expanded beyond ordinary imagination.

The universe became far larger, older, and stranger than ancient religion knew.

This discovery did not make the universe less wonderful.

It made it more wonderful.

But it also made the religious picture harder to defend.

A universe once imagined as a small human-centered stage turned out to be vast beyond human scale. If the cosmos was designed mainly for us, it is an unusual design.

## From Center to Smallness

Human beings naturally feel central to their own lives.

We wake inside our own thoughts. We see the world from our own eyes. Our pain feels urgent. Our hopes feel important. Our families and communities feel like the center of meaning.

Religion often expanded this feeling into the structure of the universe.

Human beings were placed at the center. Earth was the main stage. The heavens were arranged above. God watched human behavior. The universe existed as a moral drama about us.

Science gradually removed that centrality.

Earth is not the center of the solar system. The sun is not the center of the galaxy. The Milky Way is not the center of the universe. Human beings appeared very late in cosmic history. Our species occupies a tiny planet for a brief moment in an immense natural story.

This can feel humbling.

But humility is not despair.

To be small is not to be worthless. A child is small beside a mountain, but not meaningless. A human life is brief beside cosmic time, but not empty. Small things can matter deeply to beings who can love, suffer, remember, and care.

The problem is not that science made human life meaningless.

The problem is that science made the old human-centered religious picture less believable.

The universe does not appear arranged around us.

It appears indifferent to us.

## The Big Bang Is Not a Creation Story

Many believers point to the Big Bang and say, “Science proves creation.”

But the Big Bang is not a religious creation story.

It is a scientific model describing the early hot, dense state of the observable universe and its expansion over time. It is supported by evidence such as cosmic expansion, the cosmic microwave background, and the abundance of light elements.

It does not say that a personal God created the universe.

It does not prove scripture.

It does not prove prayer, miracles, heaven, hell, soul, divine judgment, or a cosmic plan for human beings.

The Big Bang tells us something remarkable: the universe has a deep physical history. Space, time, matter, energy, stars, galaxies, planets, chemistry, and life are part of an unfolding natural process.

What came before the earliest state we can describe? Was there a “before” in the ordinary sense? Is the universe part of a deeper physical reality? Are there laws or conditions beyond what we currently understand?

These are difficult questions.

Science does not yet have a final answer.

But “we do not yet know” does not mean “therefore God.”

The Big Bang expanded scientific understanding. It did not restore ancient theology.

## The Problem With "Everything Needs a Cause"

A common argument says:

Everything needs a cause.
The universe exists.
Therefore, the universe needs a cause.
That cause is God.

The argument sounds simple, but it has a problem.

If everything needs a cause, then God also needs a cause. If God does not need a cause, then the rule has already been changed. The believer has made an exception for God.

Why is God allowed to be uncaused, but reality itself cannot be?

The issue is not simple. Cause and time may behave differently at the deepest levels of physics. Our everyday idea of cause comes from events inside the universe. It may not apply easily to the universe as a whole.

Even if the universe has some ultimate explanation, we do not know that this explanation is a conscious being.

And even if there were a first cause, it would not prove the God of religion.

A first cause does not automatically become a loving father, a lawgiver, a prayer-answerer, a miracle-worker, or a judge of souls.

The argument moves too quickly.

It turns a mystery into a person.

That is not explanation.

It is imagination placed at the edge of knowledge.

## Fine-Tuning and Its Limits

The fine-tuning argument is more subtle.

It says the universe appears to have physical conditions that allow life. If gravity, electromagnetism, nuclear forces, particle masses, or other constants were very different, stars, planets, chemistry, and life might not exist.

This is a serious question.

The universe is not chaos. It has structure. Matter can form atoms. Atoms can form stars. Stars can form heavier elements. Chemistry can produce complexity. Under rare conditions, life can arise and evolve.

That is astonishing.

But astonishment is not proof of God.

We do not yet know whether the constants of nature could have been different. We do not know whether deeper laws explain them. We do not know whether many universes exist with different conditions. We do know that any observer must find itself in a universe compatible with observers.

Fine-tuning may point to a deep question in physics.

It does not point clearly to the personal God of religion.

There is also the scale problem.

If the universe was designed for life, most of it is strangely lifeless. Vast regions are empty. Space is cold and lethal. Radiation kills. Stars explode. Black holes consume. Most planets are barren. Even on Earth, life exists through struggle, extinction, disease, predation, and accident.

The universe permits life.

It does not look carefully arranged for human comfort.

## A Hostile and Beautiful Cosmos

The universe is beautiful, but it is not kind.

Stars shine, but they also burn out.

Galaxies are majestic, but they collide.

Planets form, but most are lifeless.

Asteroids move silently until impact.

Radiation does not care about DNA.

Gravity does not care about suffering.

Vacuum does not care about breath.

This is not an insult to the universe. It is an observation.

The cosmos is not cruel in the human sense because cruelty requires intention. But it is indifferent. It follows physical law without moral concern.

That matters for the God question.

If the universe were governed by a loving personal God, we might expect nature to show some clear moral direction. Instead, the universe operates through forces that do not recognize innocence, prayer, virtue, or need.

A newborn child and a stone are equally subject to gravity.

A saint and a criminal are equally vulnerable to radiation.

A prayerful family and a nonreligious family are equally exposed to earthquake, disease, accident, and death.

The universe behaves lawfully.

It does not behave morally.

## Natural Laws Are Not Moral Laws

Believers sometimes see natural law as evidence of divine lawgiver.

The universe has order. Therefore, they say, there must be a mind behind it.

But order alone does not prove a personal God.

Natural laws describe regular patterns in reality. They tell us how matter, energy, space, time, and forces behave. They do not tell us that the universe has moral intention.

Gravity pulls the murderer and the innocent child with equal force.

Fire burns the devoted believer and the atheist the same way.

Viruses infect according to biology, not virtue.

Earthquakes follow geology, not prayer.

Natural law is powerful because it is consistent. That consistency allows science, medicine, engineering, and prediction. But consistency is not the same as care.

A universe governed by natural law can be intelligible without being morally guided.

We can admire the order of nature without turning order into divine love.

The laws of physics are not commandments.

They do not judge. They do not forgive. They do not protect. They simply describe how nature behaves.

## "God Did It" Explains Too Little

When faced with cosmic mystery, religion often says, “God did it.”

But what does that explain?

How did God create the universe?
By what process?
Why this universe and not another?
Why so much emptiness?
Why so much danger?
Why billions of years before human beings?
Why a cosmos mostly hostile to life?
Why no clearer sign of divine purpose?

“God did it” often stops the question before it becomes scientific.

It gives a name, not a mechanism.

It gives emotional satisfaction, not understanding.

Science does not answer everything, but it asks better questions. It studies expansion, particles, fields, gravity, radiation, stars, galaxies, cosmic background light, and physical law. It tests ideas against observation. It corrects itself when evidence demands correction.

Religion may answer faster.

Science answers more carefully.

A quick answer is not always a true answer.

## Even a Creator Would Not Prove Religion

Suppose, for the sake of argument, that the universe had some unknown creator.

What would follow?

Not much of religion.

It would not prove that any scripture is divinely revealed. It would not prove that prayers are heard. It would not prove miracles. It would not prove heaven or hell. It would not prove karma, rebirth, angels, demons, prophets, incarnations, rituals, or divine judgment.

It would not even prove that the creator is loving.

A creator could be indifferent. A creator could be limited. A creator could be impersonal. A creator could be unlike anything described by religion. A creator could have made the universe and never interacted with it again.

This is why many arguments for God fail even when granted more than they deserve.

They may point to a possible unknown source of the universe, but then religion quietly adds many extra claims.

The leap from "the universe may have a cause" to "my religion is true" is enormous.

The evidence does not carry it.

## The Emotional Pull of a Created Universe

The idea of a created universe has emotional power.

It makes reality feel intentional. It makes human life feel wanted. It suggests that behind the silence of space there is mind, purpose, and care. It tells us we are not accidents in a vast cosmos, but children of a divine plan.

It is easy to understand why this comforts people.

But emotional need cannot decide cosmic truth.

Human beings may want the universe to have a parent, but wanting does not make it so. We may want death to be defeated, justice to be guaranteed, and suffering to be explained, but desire is not evidence.

A natural universe may feel emotionally difficult at first because it removes cosmic protection.

But it also removes cosmic illusion.

It allows us to see the universe as it is: vast, lawful, beautiful, dangerous, ancient, and indifferent.

That honesty can be frightening.

It can also be liberating.

## Meaning in a Natural Universe

If the universe has no divine plan, does life lose meaning?

No.

Cosmic purpose is not the only kind of meaning.

Meaning can arise within life itself. It can arise from love, friendship, work, learning, service, creativity, justice, beauty, and the reduction of suffering.

A conversation does not need to last forever to matter.

A life does not need to be written into the stars to be precious.

A kindness does not need divine approval to be good.

A child's laughter is not less real because the universe is vast.

The natural universe does not give us ready-made meaning from above. It asks us to create and protect meaning within the lives we actually have.

That is not emptiness.

It is responsibility.

The stars do not need to care about us for us to care about one another.

## The Universe as Evidence Against the Personal God

The universe is not evidence against every possible vague spiritual idea.

But it is difficult evidence for the personal God of religion.

The cosmos is too vast to look human-centered.
Too old to fit ancient creation stories.
Too hostile to look designed for our comfort.
Too silent to clearly reveal divine love.
Too indifferent to behave like moral governance.
Too natural in its operation to require constant divine action.

This does not remove wonder.

It redirects wonder.

We can stand under the night sky and feel awe without turning awe into worship. We can study galaxies without pretending they were arranged for us. We can accept our smallness without losing dignity.

The universe does not need to be a divine stage in order to be magnificent.

It is magnificent as nature.

And when nature is enough, God becomes unnecessary.

# Chapter Summary

1. Ancient religion often placed God in the sky because the heavens felt mysterious, powerful, and unreachable.
2. Modern cosmology shows a universe far larger, older, and less human-centered than ancient religion imagined.
3. The Big Bang is a scientific model, not proof of religious creation.

4. First cause and fine-tuning arguments raise real questions, but they do not prove the personal God of religion.
5. The universe is lawful and beautiful, but also vast, hostile, silent, and morally indifferent.
6. Natural laws describe how reality behaves. They do not show divine care or moral judgment.
7. Meaning can exist within a natural universe without being imposed from above.

# The Case So Far

The previous chapter showed how science repeatedly replaced supernatural explanations within the world. This chapter turned to the universe as a whole and found the same pattern: cosmic mystery remains, but God is not required as an explanation.

# Key Takeaway

The universe inspires awe, but awe is not evidence for God. The cosmos is vast, ancient, lawful, beautiful, and dangerous. It permits life, but it does not appear centered on human beings or governed by moral care. The universe is better understood as natural reality than as evidence for a personal God.

# Sources and Further Reading

1. Carl Sagan, **Cosmos**.
2. Carl Sagan, **Pale Blue Dot**.
3. Stephen Hawking, **A Brief History of Time**.
4. Stephen Hawking and Leonard Mlodinow, **The Grand Design**.

5. Steven Weinberg, **The First Three Minutes**.
6. Sean Carroll, **The Big Picture**.
7. Lawrence M. Krauss, **A Universe from Nothing**.

## Transition to the Next Chapter

The universe does not clearly reveal a divine hand. But many believers turn from the stars to living things and say, "Life itself proves design." The next chapter examines that claim and asks whether life truly needs a designer, or whether evolution explains what design once seemed to require.

# Chapter 6: Life Without a Designer

## Why Life Looks Designed

Life is astonishing.

A bird lifts itself into the air. A flower opens toward light. A seed becomes a tree. A spider builds a web. A baby recognizes a mother's voice. A heart begins beating before birth. A wound heals. A cell divides. A forest renews itself after fire.

It is easy to understand why human beings saw design in life.

Living things appear purposeful. Eyes see. Wings fly. Roots absorb water. Lungs breathe. Blood carries oxygen. Bees find flowers. Plants turn sunlight into food. Bodies grow, repair, reproduce, and defend themselves.

Before modern biology, this looked like direct evidence of a designer.

If a watch has a watchmaker, many argued, then living organisms must have a divine maker.

But the comparison is misleading.

A watch does not reproduce. It does not vary across generations. It does not compete, adapt, mutate, inherit traits, or evolve. Living things do.

Life looks designed because natural selection can produce the appearance of design.

This was one of the greatest discoveries in human history.

## Darwin's Great Insight

Charles Darwin did not remove wonder from life.

He explained why life could become so wonderfully adapted without being planned.

His central insight was simple but powerful.

Living things vary. Some variations help survival and reproduction. Organisms with helpful traits are more likely to leave offspring. Those offspring inherit traits. Over long periods, small differences can accumulate. Gradually, populations change.

This is natural selection.

It is not pure chance.

Chance produces variation. Selection preserves useful variation. Over enough time, this process can shape organs, instincts, bodies, colors, behaviors, and relationships between organisms and environments.

No conscious designer is needed.

A fish does not need to know it is becoming better suited to water. A bird does not need to know its wing is being shaped. A flower does not need to know it is attracting pollinators. The process has no foresight, but it has consequences.

Those consequences can be remarkable.

Darwin showed that life could have a natural history.

That changed everything.

## Evolution Is Not "Just Chance"

One of the most common misunderstandings about evolution is that it says life came about by chance.

That is false.

Evolution includes chance, but it is not only chance.

Mutations may occur without purpose. Genetic variation may arise through processes that do not aim at a goal. But natural selection is not random in the same way. Traits that help organisms survive and reproduce tend to be preserved more often than traits that do not.

This makes evolution powerful.

Imagine a river shaping stone. No mind directs each drop of water. Yet over long time, flowing water can carve valleys, smooth rocks, and change landscapes. Natural processes can create structure without intention.

Evolution works in a similar natural way.

It does not need to plan the future. It only needs variation, inheritance, reproduction, and selection across deep time.

The result can look designed.

But looking designed is not the same as being designed.

## The Tree of Life

Evolution does not merely explain adaptation.

It explains relationship.

All living things are connected through common ancestry. Humans, birds, fish, insects, trees, bacteria, and fungi are not separate creations placed independently on Earth. They are branches of a vast tree of life.

This idea is supported by many lines of evidence.

Fossils show change across time.

Comparative anatomy shows shared structures. The bones in a human arm, a bat wing, a whale flipper, and a cat's forelimb follow related patterns because they come from common ancestry.

Embryology shows developmental similarities among related organisms.

DNA gives even deeper evidence. Living things carry molecular records of relationship. Species that are closely related share more genetic similarity. Humans share common ancestry with other primates. All life uses the same basic genetic language, with variations built on shared biological foundations.

This is exactly what evolution predicts.

It is not what we would expect from separate acts of perfect design.

Life looks less like isolated inventions and more like a branching family history.

## DNA and the Record of Descent

DNA is often presented as evidence for God because it contains biological information.

But DNA is better understood as evidence for evolution.

DNA carries instructions for building and regulating living organisms, but it also carries history. It contains mutations, duplications, inherited errors, inactive genes, viral remnants, and patterns of similarity that reveal ancestry.

If life were separately designed from scratch, we might expect clean independent designs.

Instead, we find shared genetic patterns.

We find organisms carrying marks of common descent. We find broken genes shared by related species. We find molecular similarities that match the branching pattern already suggested by fossils and anatomy.

DNA does not look like a perfect code written fresh for each species.

It looks like a long, edited, copied, damaged, repaired, duplicated, and modified biological record.

That is exactly what evolution would produce.

The deeper we look into life, the more historical it becomes.

## Imperfection Is Evidence

A perfect designer could design from scratch.

Evolution cannot.

Evolution must work with what already exists. It modifies old structures. It reuses parts. It builds through compromise. It does not start over with a clean plan.

That is why living things often show imperfect arrangements.

Some features are useful but inefficient. Some are leftovers from earlier forms. Some create vulnerability. Some work well enough for survival but not well enough to avoid pain, disease, or failure.

This matters.

Imperfection is not a small embarrassment for design. It is one of the strongest clues to evolution.

A body shaped by history will carry marks of history.

A body designed perfectly from nothing should not.

This does not mean life is poorly made in every way. Life is often astonishingly effective. But its effectiveness is mixed with compromise, waste, and vulnerability.

That mixture is exactly what we expect from evolution.

It is much harder to explain through perfect design.

## Extinction and Waste

Evolution also explains extinction.

Most species that have ever lived are gone. Earth's history is filled with life forms that appeared, struggled, reproduced, changed, and disappeared. Entire branches of life ended. Mass extinctions destroyed countless organisms. Environments shifted. Competition changed. Asteroids struck. Climate changed. Species failed.

This is difficult to reconcile with the idea of a careful designer creating each species for a clear purpose.

Why create so much life only to let most of it vanish?

Why fill Earth's history with suffering, struggle, predation, and extinction long before human beings appeared?

Evolution makes sense of this.

Life is not a finished plan. It is a branching process. Some branches flourish for a time. Some fail. Some adapt. Some disappear. There is no guarantee of permanence. There is no built-in protection from extinction.

Nature does not preserve species because they are beautiful, innocent, or divinely intended.

It preserves them only if they survive and reproduce under changing conditions.

Extinction is not a problem for evolution.

It is part of the story.

## Disease and Biological Struggle

Life is not only beautiful.

It is also full of conflict.

Viruses invade cells. Bacteria multiply. Parasites feed on hosts. Predators kill prey. Cancer grows from the body's own cells. Genetic mutations cause suffering. Immune systems defend, but sometimes attack the body itself. Reproduction is often risky. Aging slowly weakens living tissues.

A perfectly loving designer raises difficult questions here.

Why design a world where organisms live by consuming one another? Why create parasites that blind children, insects that spread deadly disease, bacteria that kill infants, and viruses that exploit the machinery of life?

Evolution gives a clearer answer.

Living things are not designed for universal happiness. They are shaped by survival and reproduction. A virus succeeds by copying itself. A parasite succeeds by using a host. A predator succeeds by catching prey. A cancer cell grows because cellular control fails.

This is not moral.

It is biological.

Evolution explains why life contains both beauty and cruelty. It explains why organisms are adapted not for kindness, but for survival in competition with other organisms.

A divine design of life struggles to explain this.

Evolution expects it.

## The Failure of "Irreducible Complexity"

Some defenders of design argue that certain biological systems are too complex to have evolved gradually. They say such systems could not work if any part were missing, so they must have been created all at once.

This is often called irreducible complexity.

The argument sounds strong, but it misunderstands evolution.

Structures do not need to appear in their final form all at once. Parts can have earlier functions. A structure used for one purpose can later be modified for another. Simple systems can become more complex over time. Natural selection can build step by step, using available material.

Evolution does not require a full modern eye to appear suddenly. Light-sensitive cells can be useful before a complex eye exists. A shallow cup of light-sensitive cells can give direction. A deeper cup can improve focus. A lens can refine vision. Each step can be useful.

The same principle applies broadly.

Biological systems have histories.

Complexity can be built gradually when each stage gives some advantage, or when old parts are reused in new ways.

Irreducible complexity often reflects our lack of imagination about natural pathways, not a failure of evolution.

## Human Beings Are Part of Nature

Many people resist evolution because it includes human beings.

They may accept that animals evolved, but feel that humans must be separate. We think, love, speak, create, mourn, reason, and ask about meaning. Surely, some say, we must be more than evolved animals.

Human beings are indeed remarkable.

But remarkable does not mean supernatural.

We share anatomy, genetics, development, emotions, instincts, and ancestry with other life. Our hands, eyes, spine, teeth, blood, brain, immune system, reproductive system, and DNA all place us within the natural world.

Evolution does not say human beings are "just animals" in a dismissive sense.

It says we are animals with unusual capacities: language, memory, moral reflection, imagination, science, art, medicine, and self-awareness.

A natural origin does not remove dignity.

A child is not less precious because reproduction is biological. Love is not less real because it depends on the brain. Human life is not less valuable because it evolved.

Evolution explains our origin.

It does not erase our worth.

## Evolution and Morality

Some people fear that if evolution shaped human beings, morality becomes meaningless.

But this fear is misplaced.

Evolution helps explain why moral instincts could arise. Social animals benefit from care, cooperation, attachment, trust, fairness, and protection of offspring. Human beings, with language and memory, developed these capacities further into moral systems, laws, duties, and ideals.

This does not mean every evolved impulse is good.

Evolution also gave us fear, aggression, tribalism, selfishness, jealousy, and cruelty. Nature is not a perfect moral guide.

That is why morality requires more than instinct.

It requires reason, empathy, education, law, reflection, and the widening of concern beyond tribe, religion, caste, nation, or family.

Evolution gives the roots of moral life.

Human responsibility must cultivate the tree.

God is not needed to explain why social beings develop moral concern.

## Why Evolution Supports Atheism

Evolution does not logically disprove every possible God.

A believer can say that God used evolution.

But this move changes the original design claim.

Evolution shows that life does not require special creation. It explains complexity without a designer. It explains common ancestry without separate divine acts. It explains imperfection, extinction, disease, and biological struggle better than perfect design.

A God who uses billions of years of mutation, suffering, predation, extinction, genetic disease, parasites, and death to produce life is very different from the loving designer many religions imagine.

Such a God becomes harder to distinguish from no God at all.

The more evolution explains, the less work God has left to do.

That is why evolution is central to science-based atheism.

It removes one of the oldest and strongest arguments for God: the appearance of design in life.

## The Beauty of Undesigned Life

Some people think that if life is not designed, it becomes less beautiful.

The opposite can be true.

Life becomes more beautiful when we understand its deep history.

A bird is not merely a crafted object. It is the living result of ancient ancestry, selection, bone, feather, muscle, instinct, migration, and time.

A flower is not merely decoration. It is part of a long relationship among plants, insects, sunlight, soil, color, scent, and reproduction.

A human hand is not merely a tool. It carries the history of fins, limbs, grasping, touch, work, art, surgery, writing, and care.

Evolution does not flatten wonder.

It deepens it.

We are not separate from life. We are part of its long story. We carry ancient history in our bones, blood, genes, and brains.

Life did not need to be designed to be magnificent.

It is magnificent because nature, across deep time, produced forms capable of movement, sensation, memory, love, suffering, and thought.

That is wonder enough.

# Chapter Summary

1. Life looks designed because organisms are complex and adapted to their environments.
2. Darwin showed how natural selection can produce the appearance of design without a designer.
3. Evolution is not pure chance. Variation may arise without purpose, but selection preserves useful traits.
4. Fossils, anatomy, embryology, and DNA support common ancestry.

5. Imperfection, extinction, disease, parasites, and biological struggle fit evolution better than perfect design.
6. Human beings are part of nature, and a natural origin does not remove human dignity.
7. Evolution removes the need for special creation while preserving the wonder of life.

## The Case So Far

The previous chapter showed that the universe does not clearly reveal a divine hand. This chapter turned to life and showed that its beauty, complexity, and adaptation do not require a designer. Evolution explains life as a natural, historical process.

## Key Takeaway

Life appears designed, but evolution explains that appearance through natural selection, common ancestry, variation, inheritance, and deep time. The living world is beautiful, but also imperfect, wasteful, painful, and historical. Life does not need a divine designer to be understood or admired.

## Sources and Further Reading


1. Charles Darwin, **On the Origin of Species**.
2. Richard Dawkins, **The Blind Watchmaker**.
3. Jerry A. Coyne, **Why Evolution Is True**.
4. Ernst Mayr, **What Evolution Is**.
5. Neil Shubin, **Your Inner Fish**.
6. Sean B. Carroll, **Endless Forms Most Beautiful**.

7. Douglas J. Futuyma and Mark Kirkpatrick, **Evolution**.

## Transition to the Next Chapter

Evolution explains life in general. But the argument becomes even more personal when we look at our own bodies. If human beings were specially designed by a perfect creator, what should we expect to find in the body? The next chapter examines the human body itself: brilliant, fragile, and far from perfect.

# Chapter 7: The Body: Brilliant, Fragile, and Poorly Designed

## The Body That Amazes and Fails

The human body is extraordinary.

The heart beats thousands of times each day. The lungs move air in and out without conscious effort. Blood carries oxygen to every living tissue. The kidneys filter. The liver processes. The immune system defends. The brain receives light, sound, pain, memory, language, emotion, and thought. A single fertilized cell becomes a newborn child.

As a physician, I have never lost my sense of wonder at the body.

But wonder is not the same as evidence of perfect design.

The same body that amazes us also fails in painful and predictable ways. The spine bends and aches. The throat can choke. The pelvis can make childbirth dangerous. Cells can become cancer. Genes can carry devastating mutations. The immune system can attack the body it is meant to defend. The brain can lose memory, speech, judgment, and personality. Aging slowly weakens almost every organ.

The human body is not a simple argument for God.

It is an argument for evolution, compromise, fragility, and natural history.

If the body were designed from scratch by a perfect creator, we might expect cleaner engineering. Instead, we find brilliance mixed with weakness, efficiency mixed with danger, and function mixed with failure.

That is exactly what evolution predicts.

## Evolution Builds With What Already Exists

A perfect engineer can begin with a clean design.

Evolution cannot.

Evolution works with existing structures. It modifies what is already present. It does not begin again from the beginning. It preserves what helps survival and reproduction, even if the result is awkward, painful, or vulnerable later in life.

This is why the body often looks less like a machine designed from scratch and more like an old house repaired again and again.

A wall is moved. A pipe is redirected. A new room is added. Wires are extended. The house works, but the layout carries the marks of its history.

The body is similar.

It carries ancestry. It carries compromise. It carries traces of older forms. It solves one problem while creating another. It is good enough to survive and reproduce, not perfect enough to avoid all suffering.

This is the key to understanding human anatomy.

The body is not poorly made in every way. It is often astonishingly effective. But it is not perfect. It is historical.

## The Spine and the Price of Standing Upright

The human spine is one of the clearest examples of evolutionary compromise.

Our ancestors moved on four limbs. Over time, human beings became upright walkers. This freed the hands for carrying, tool use, work, care, and communication. It helped make human life possible in the form we know.

But standing upright came with a price.

The spine now carries the weight of the upper body vertically. The lower back bears enormous stress. Discs can bulge or rupture. Nerves can be compressed. Curves in the spine help balance the body, but they also create strain. Back pain is one of the most common human complaints.

A perfect designer could have built a spine specifically for upright life.

Evolution had to modify an existing structure.

The result works well enough for walking, running, lifting, bending, and carrying. But it also produces chronic pain, disability, surgery, and suffering.

The spine is useful.

It is not ideal engineering.

## Childbirth and the Human Pelvis

Human childbirth is another powerful example.

The human brain became large. The human pelvis, shaped by upright walking, had limits. The baby's head became large relative to the birth canal. The result is one of the most difficult childbirth processes in the animal world.

Before modern medicine, childbirth was a major cause of death for mothers and babies. Even today, childbirth can become dangerous without skilled care, surgery, blood transfusion, antibiotics, anesthesia, and emergency obstetric support.

Why would a perfect designer create such a narrow passage for such a large head?

Evolution explains the problem.

The pelvis had to balance two demands: walking upright and giving birth to large-brained infants. Natural selection did not create a perfect solution. It produced a compromise.

That compromise allowed human beings to survive as a species.

But it also brought pain, risk, injury, and death.

The delivery room is not strong evidence of perfect design.

It is strong evidence of biological trade-off.

## The Throat That Can Choke

Human beings breathe and swallow through a shared passage.

This arrangement allows speech, eating, breathing, and complex vocal control. It also creates danger. Food can enter the airway. A person can choke. A small object can block breathing. A meal can become fatal within minutes.

This is not a rare theoretical problem.

Choking is a real cause of injury and death.

A perfect designer could have separated the airway and food passage more safely. Evolution inherited structures from earlier animals and modified them. The result gives us speech and swallowing, but also risk.

Again, the pattern is compromise.

One advantage comes with another danger.

The human throat is not stupid. It works most of the time. But it is not the kind of design one would expect from a flawless engineer starting from nothing.

It is the kind of arrangement one expects from evolution.

## The Eye and the Blind Spot

The human eye is often used as evidence for design.

And the eye is indeed remarkable. It focuses light, adjusts to brightness, detects color, tracks motion, and sends information to the brain with astonishing speed.

But the eye also shows imperfection.

In the human retina, nerve fibers and blood vessels lie in front of the light-sensitive cells. The nerve fibers must pass through the retina to leave the eye, creating a blind spot. The brain fills in the missing area so smoothly that most people never notice it.

But the blind spot is real.

This arrangement is not what a perfect engineer would necessarily choose. It is what one might expect from a structure shaped through evolutionary history.

The eye works well.

But working well is not the same as perfect design.

Evolution can produce excellent function through imperfect pathways.

## The Appendix, Wisdom Teeth, and Leftover Problems

The body also contains structures that make sense through ancestry.

The appendix may have some immune-related functions, but it can also become inflamed, rupture, and kill without surgery or antibiotics. Wisdom teeth may have been more useful in ancestors with different diets, larger jaws, and heavier tooth wear. In modern human jaws, they often become impacted, painful, infected, or require removal.

These are not the strongest arguments by themselves, but they fit the larger pattern.

The body carries leftovers, reduced functions, old structures, and changing uses.

Evolution does not erase every unnecessary or risky part. If a structure does not prevent survival and reproduction too strongly, it may persist. If it causes problems mostly later in life, after reproduction, selection may not remove it efficiently.

This helps explain why the body contains so many imperfections.

It is not a perfect plan.

It is a history.

# Cancer: The Body Against Itself

Cancer is one of the most powerful challenges to the idea of perfect design.

Cancer is not an outside enemy in the same way as a virus or bacterium. It arises from the body's own cells. Normal cells must divide, repair, respond to signals, and sometimes die when damaged. These processes are essential for life.

But the same processes can fail.

DNA can mutate. Repair systems can miss errors. Cells can escape control. Growth signals can become abnormal. Cells can invade, spread, and destroy the body that gave them life.

In cancer, life turns against itself.

This is deeply difficult to reconcile with the idea of a body perfectly designed by a loving creator. Why make life depend on cell division while allowing that same process to produce tumors in children, adults, and the elderly? Why allow the machinery of growth to become the machinery of suffering?

Evolution explains the vulnerability.

Bodies are made of cells that must divide. DNA copying is powerful but not perfect. Natural selection works strongly on survival and reproduction, but not toward eternal health. As organisms live longer, mutations accumulate. Some cells gain growth advantages inside the body, much as organisms compete in nature.

Cancer is not moral punishment.

It is biology.

That truth matters.

It removes blame and directs us toward research, prevention, screening, treatment, and compassion.

## Genetic Disease and Innocent Suffering

Genes carry the instructions and regulatory patterns of life.

They also carry risk.

A single mutation can cause severe disease. A child can be born with a genetic disorder before any moral choice is possible. Families can carry inherited conditions for generations. Some diseases damage the brain, blood, lungs, muscles, nerves, or metabolism from early life.

Nothing about this looks like justice.

A newborn has not sinned. A child has not chosen a fate. A gene does not know virtue. A mutation does not respect innocence.

Religious explanations sometimes try to give such suffering a spiritual meaning. But as a physician, I find the biological explanation both truer and more humane.

Genetic disease follows inheritance, mutation, cell biology, and development.

Not sin.

Not divine punishment.

Not moral desert.

Once we understand this, our response improves. We do not blame the child. We study the disease. We support the family. We seek treatment. We develop genetic counseling, newborn screening, molecular medicine, and better care.

Natural explanation protects the innocent from theological blame.

## The Immune System: Defender and Destroyer

The immune system is one of the body's great marvels.

It recognizes threats, remembers past infections, coordinates cells, produces antibodies, and protects us from countless dangers. Without it, ordinary life would be impossible.

But the immune system is also dangerous.

It can overreact in allergy. It can attack the body in autoimmune disease. It can inflame tissues, damage organs, and contribute to suffering. In some infections, the immune response itself can become part of the harm.

This is not perfect defense.

It is a powerful system with risks.

Evolution shaped immunity in a world full of infection, injury, parasites, and environmental danger. A strong immune response could save life. But strength can become misdirection. Defense can become destruction.

This is another pattern of compromise.

The immune system is not useless. It is essential. But it is not perfectly designed for painless human well-being.

It is an evolved defense system, magnificent and dangerous at the same time.

## Aging and the Slow Failure of the Body

Aging is sometimes treated as natural in a comforting sense.

But medically, aging is also progressive vulnerability.

Skin thins. Arteries stiffen. Bones weaken. Muscles shrink. Vision dims. Hearing declines. Kidneys lose reserve. The immune system changes. The brain becomes more vulnerable. Cancer risk rises. Healing slows. Falls become more dangerous. Dementia becomes more common.

If the body were designed by a perfect creator, why would it be built to decline in this way?

Evolution helps explain it.

Natural selection is strongest before and during reproductive years. It is less powerful in removing problems that appear late in life. Traits that help early survival and reproduction may remain even if they contribute to late-life disease.

The result is a body capable of extraordinary life, but not indefinite preservation.

Aging is not a divine lesson written into flesh.

It is biological wear, imperfect repair, accumulated damage, genetic trade-offs, and the limits of evolved systems.

Understanding aging naturally does not remove dignity from the elderly.

It gives us better reasons to care for them.

## The Brain: The Most Fragile Treasure

The brain is the organ through which we experience the world.

Memory, personality, speech, love, fear, judgment, imagination, and self-awareness depend on it. It is the most precious tissue in the body because damage to it can change the person we know.

But the brain is also fragile.

A small stroke can remove speech. A head injury can alter personality. Lack of oxygen can destroy consciousness. Dementia can slowly erase memory. Infection, tumors, seizures, drugs, metabolic problems, and trauma can change thought and behavior.

If an immortal soul were the true independent self, the brain's power over personality and identity would be difficult to understand. But if the mind depends on the brain, these facts make sense.

The brain is not evidence of a soul placed inside a body.

It is evidence that the self is deeply biological.

This does not make human beings less valuable.

It makes the protection of the brain, the care of patients, and the dignity of the vulnerable even more important.

## The Physician's View of Design

A physician sees the body from a different angle than a person admiring anatomy from a distance.

In the clinic, emergency room, hospital ward, or intensive care unit, the body is not an abstract marvel. It is a living system under stress.

The heart fails. The lungs stiffen. The kidneys shut down. Blood clots. Cells mutate. Infections spread. Arteries block. Nerves degenerate. Bones fracture. Memory disappears. Pain overwhelms. Breathing becomes work.

Medicine exists because the body fails.

Hospitals exist because the body fails.

Surgery exists because the body fails.

Public health exists because the body fails and because nature does not protect us reliably.

This does not make the body ugly. It makes it honest.

The body is wonderful, but it is not perfect. It is resilient, but not secure. It is organized, but not flawless. It is capable of healing, but also of collapse.

The physician's view does not remove wonder.

It removes illusion.

## The Selective Eye of the Design Argument

The design argument often chooses only the body's successes.

It points to the eye, but not the blind spot.

It points to the heart, but not congenital heart disease.

It points to childbirth, but not maternal death.

It points to the immune system, but not autoimmune disease.

It points to the brain, but not dementia.

It points to healing, but not cancer.

It points to life, but not genetic suffering.

This selective view is not fair.

A serious explanation must account for both success and failure.

Evolution does.

Design struggles.

Evolution expects useful structures mixed with imperfection. It expects trade-offs. It expects disease. It expects aging. It expects inherited limitations. It expects systems that work well enough, not systems designed for perfect health and happiness.

The body's brilliance is real.

So is its fragility.

A complete explanation must include both.

## The Body Is Precious Because It Is Fragile

Some people fear that if the body is evolved, it becomes less sacred.

I see the opposite.

The body is precious because it is fragile.

Health is precious because it can be lost.

Memory is precious because it can fade.

Breath is precious because it can fail.

Life is precious because it is temporary.

A natural body does not deserve less care. It deserves more. There is no need to imagine divine design in order to value human life. The patient in pain matters because pain is real. The child with disease matters because the child can suffer. The elderly person with dementia matters because personhood, memory, vulnerability, and love matter.

The body does not need to be perfect to be worthy of care.

It needs only to be the living home of a conscious being.

That is enough.

## No Perfect Designer Required

The human body is one of the strongest reasons many people believe in God.

But when examined carefully, it points in another direction.

It is brilliant, but flawed.

It is functional, but full of compromise.

It is beautiful, but vulnerable.

It is capable of healing, but also of disease.

It carries ancestry, trade-offs, mutations, leftovers, and limits.

This is not what we would expect from perfect design.

It is what we would expect from evolution.

The body does not need a divine engineer to be understood. It needs biology, anatomy, physiology, genetics, evolution, medicine, and compassion.

And once we understand it naturally, we can care for it more wisely.

# Chapter Summary

1. The human body is extraordinary, but not perfectly designed.
2. Evolution builds with existing structures and produces compromise rather than flawless engineering.
3. The spine, pelvis, throat, eye, appendix, wisdom teeth, immune system, brain, and aging body all show trade-offs and vulnerabilities.
4. Cancer and genetic disease are especially powerful challenges to the idea of a perfectly designed body.
5. Disease follows biology, not sin or divine punishment.
6. Medicine exists because the body is fragile and nature does not reliably protect us.
7. The body is precious not because it is perfect, but because it is the fragile home of conscious life.

## The Case So Far

The previous chapter showed that evolution explains life without a designer. This chapter brought that argument into the human body itself. The body is wonderful, but its flaws, diseases, compromises, and failures make far more sense as products of evolution than as perfect design.

## Key Takeaway

The human body is not evidence of flawless divine engineering. It is evidence of evolutionary history: useful structures mixed with weakness, beauty mixed with disease, resilience mixed with failure. The body is precious, but it is not perfect. It needs care, medicine, and compassion, not supernatural explanation.

## Sources and Further Reading


1. Charles Darwin, **On the Origin of Species**.
2. Neil Shubin, **Your Inner Fish**.
3. Randolph M. Nesse and George C. Williams, **Why We Get Sick**.
4. Randolph M. Nesse, **Good Reasons for Bad Feelings**.
5. Siddhartha Mukherjee, **The Emperor of All Maladies**.
6. Siddhartha Mukherjee, **The Gene**.
7. Jerry A. Coyne, **Why Evolution Is True**.


## Transition to the Next Chapter

The body and the living world show deep evidence of evolution. But another question remains: how did life begin in the first place? Many believers place God at that earliest boundary. The next chapter examines the origin of life and asks whether an unanswered scientific question should be treated as evidence for God.

# Chapter 8: The Origin of Life and the God of the Gaps

## The Question Before Evolution

Evolution explains how life changes after life already exists.

But many believers ask a different question:

How did life begin in the first place?

This is an important question. It should not be dismissed. A living cell is not simple. Even the simplest life must do remarkable things. It must maintain some boundary between itself and the environment. It must use energy. It must carry information. It must reproduce with variation. It must preserve enough stability to continue, while allowing enough change to evolve.

That is not a small problem.

The origin of life is one of the great scientific questions.

But a difficult question is not automatically evidence for God.

This chapter is about the boundary between honest mystery and premature religious certainty. It asks whether the origin of life requires divine creation, or whether it remains a natural question still under investigation.

The answer matters because religion often lives at the edge of what science has not yet fully explained.

That edge is where the God of the gaps appears.

## Abiogenesis Is Not Evolution

A common confusion must be cleared first.

Evolution and the origin of life are not the same question.

Evolution explains how living populations change across generations once reproduction, inheritance, variation, and selection exist. It explains common ancestry, adaptation, biological diversity, imperfection, and extinction.

The origin of life, often called abiogenesis, asks how the first life-like systems arose from nonliving chemistry.

These are related subjects, but they are not identical.

Even if scientists did not yet know exactly how life began, evolution would still explain the history and diversity of life after it began. The evidence for evolution does not collapse because the first steps of life remain difficult.

This is important because some religious arguments confuse the two.

They say, "Science cannot explain how life began, therefore evolution is false."

That does not follow.

The origin of life is a separate scientific frontier.

Evolution remains one of the strongest explanations in all of biology.

## What Early Life Needed

To understand the origin of life, we must avoid imagining that the first living system was like a modern cell.

Modern cells are highly complex. They contain DNA, RNA, proteins, membranes, enzymes, energy systems, repair mechanisms, and regulatory networks. They are the result of billions of years of evolution.

The first life-like systems were almost certainly much simpler.

They may not have looked like modern cells. They may have been chemical systems capable of storing information, copying imperfectly, using energy, and becoming enclosed in simple membranes. Once such systems existed, natural selection could begin to act.

The key challenge is to understand how chemistry crossed into biology.

How did molecules become organized enough to copy, vary, and persist?
How did energy flow through early systems?
How did simple boundaries form?
How did early information and metabolism develop?
How did chemical networks become capable of evolution?

These are hard questions.

But they are scientific questions.

They concern chemistry, energy, geology, water, minerals, molecules, membranes, and time.

They do not require us to begin with a miracle.

## Several Natural Pathways Are Being Studied

Scientists have proposed several possible pathways for the origin of life.

Some focus on RNA, a molecule that can carry information and also help chemical reactions. This has led to the idea of an early "RNA world," where RNA-like molecules played roles later divided between DNA and proteins.

Others focus on metabolism first. They ask whether networks of chemical reactions could have begun using energy before modern genetic systems existed.

Some researchers study hydrothermal vents, where heat, minerals, chemical gradients, and ocean water may have created conditions favorable to early life-like chemistry.

Others study simple membranes and protocells. Fatty molecules can form bubbles or compartments, creating primitive boundaries that separate inside from outside.

Still others examine mineral surfaces, clay, cycles of wetting and drying, ultraviolet light, volcanic settings, and early Earth chemistry.

No single pathway is yet proven as the full answer.

But that is not a weakness of science. That is how science works. It explores, tests, revises, rejects, and improves possibilities.

The important point is this: scientists are not merely saying, "Life happened somehow." They are studying real mechanisms by which chemistry could gradually become biological.

## Why "God Did It" Does Not Explain Life

When faced with the origin of life, religion often says, "God created life."

But what does that explain?

How did God create life?
By what process?
Why use chemistry at all?
Why begin with simple life and then allow billions of

years of evolution?
Why use mutation, extinction, predation, parasites, and genetic disease?
Why create life through a history so full of waste and suffering?

“God did it” may satisfy emotion, but it does not provide a mechanism.

It does not explain the chemistry.

It does not explain the steps.

It does not explain why life looks historical, related, imperfect, and evolved.

A true explanation should increase understanding. It should tell us something more than “a hidden power chose it.”

Science may not yet have the complete answer, but it asks questions that can be studied. It searches for mechanisms. It tests conditions. It compares molecules. It looks at early Earth environments. It builds models and experiments.

“God did it” stops inquiry too soon.

## The God of the Gaps Problem

The God of the gaps is the habit of placing God wherever human knowledge is incomplete.

Once God was placed in thunder.

Then thunder became sound from lightning.

God was placed in disease.

Then disease became biology, infection, genetics, and physiology.

God was placed in life's complexity.

Then evolution explained adaptation and common ancestry.

Now God is often placed at the origin of life.

But history warns us to be careful.

A gap in scientific knowledge is not evidence for God. It is only a gap in scientific knowledge.

If we do not yet know exactly how life began, the honest answer is:

We do not yet know.

That answer may feel less satisfying than a religious story, but it is more truthful.

The God of the gaps has a long record of retreat.

As knowledge grows, the space reserved for supernatural explanation becomes smaller. Building faith on what science has not yet explained is risky because science continues to investigate.

A mystery today may become a textbook chapter tomorrow.

## Complexity Does Not Require Instant Creation

Another common mistake is to look at modern life and assume it must have appeared all at once.

A modern cell is incredibly complex. But origin-of-life research does not claim that a modern cell suddenly assembled by chance from random chemicals.

That would indeed be extremely unlikely.

The more reasonable question is whether simpler systems could have formed gradually, with each stage making the next stage more likely.

Simple molecules can form under natural conditions. Some molecules can join into larger structures. Some can help reactions. Some can form membranes. Some can store information. Some can copy imperfectly. Some chemical systems can become more stable or more persistent under certain conditions.

Once imperfect copying and variation exist, selection can begin.

The first life did not need to be as complex as a modern bacterium. It only needed to cross a threshold where chemical systems could preserve and modify themselves across time.

From there, evolution could build complexity.

This is the key point.

Modern complexity does not require instant creation.

It can arise through stages.

## Probability Arguments Are Often Misleading

Religious arguments often say that life is too improbable to arise naturally.

These arguments usually imagine modern life appearing suddenly by pure chance. They compare the origin of life to a tornado assembling an airplane or random letters producing a book.

But that is not how natural processes work.

Life did not need to appear as a complete modern cell in one step. Chemistry is not random in the way shuffled letters are random. Molecules have properties. They attract, repel, bond, fold, break, and react according to physical laws. Environments can concentrate molecules, cycle conditions, provide energy, and favor certain pathways.

Natural processes can build order step by step.

Snowflakes form patterns without design. Crystals grow. Rivers carve valleys. Stars form from gas. Planets form from disks of material. Complex structure can arise from natural law without intention.

Life is far more complex than these examples, but the principle matters.

Natural order does not automatically require a mind.

Probability arguments often fail because they assume the wrong process.

They attack sudden random assembly, not gradual chemical development under natural conditions.

## Even If Life's Origin Remained Unsolved

Suppose science never fully explained the origin of life.

What would follow?

Not God.

An unsolved problem would remain an unsolved problem.

It would not prove divine creation. It would not prove a personal God. It would not prove scripture, prayer, miracles, heaven, hell, soul, judgment, or divine love. It would not tell us which religion is true.

At most, it would show that one scientific question remains difficult.

This distinction is essential.

The failure to know a natural explanation is not the same as evidence for a supernatural explanation.

If a crime remains unsolved, we do not conclude that a ghost committed it. If a disease is not yet understood, we do not conclude that demons caused it. If a cosmic mystery remains, we do not automatically fill it with God.

Ignorance cannot carry the weight of theology.

A gap is not a foundation.

## The Origin of Life Does Not Prove a Personal God

The origin of life is sometimes treated as if it supports religion in general.

But even if someone believed life required some unknown cause, that would still not prove the personal God of religion.

It would not prove that God answers prayer.

It would not prove that God revealed any sacred text.

It would not prove that God cares about rituals, belief, worship, sin, or salvation.

It would not prove that human beings have immortal souls.

It would not prove that suffering has divine meaning.

It would not prove that one religion is true and others are false.

This is a recurring problem in arguments for God.

They begin with a mystery and end with a full religion.

The steps in between are missing.

The origin of life is a scientific problem about chemistry becoming biology. Turning that problem into proof of a personal God requires far more evidence than religion provides.

## A Better Kind of Wonder

Some people fear that a natural origin of life would make life less special.

But understanding life naturally can deepen wonder.

Imagine early Earth: oceans, heat, rock, lightning, minerals, sunlight, volcanic activity, chemical cycles, and time. Out of such conditions, simple molecules may have formed more complex systems. Some systems persisted. Some copied. Some changed. Some crossed the threshold into life.

From those beginnings came bacteria, forests, insects, fish, reptiles, birds, mammals, primates, and human beings capable of asking how it all began.

That story does not need to be supernatural to be magnificent.

In fact, its naturalness may make it more astonishing.

Life is not a magic trick placed into the world from outside. It is a deep expression of what matter and energy can become under the right conditions over time.

To understand that is not to reduce life.

It is to see life as part of the universe itself.

## Humility Without Surrender

The honest position on the origin of life is neither dogmatic religion nor arrogant certainty.

It is humility.

We do not yet know every step.

Scientists have promising ideas, but no final complete account. Some proposals may be wrong. Some may be partly right. New discoveries may change the picture.

That is acceptable.

Honest uncertainty is not weakness.

The weakness lies in pretending to know what we do not know.

Religion often rushes to fill uncertainty with certainty. Science can wait. It can say, "We do not know yet, but we can investigate."

That is a more reliable path.

The origin of life remains one of the great questions. But it does not rescue God from the retreat of supernatural explanation.

It shows only that science still has work to do.

And science is willing to do that work.

## Chapter Summary

1. Evolution and the origin of life are related but different questions.
2. Evolution explains how life changes after life exists; abiogenesis studies how life-like systems may have arisen from chemistry.
3. Scientists are investigating several natural pathways, including RNA, metabolism, hydrothermal vents, membranes, minerals, and early Earth chemistry.
4. "God did it" does not explain the mechanism by which life began.
5. The God of the gaps is weak reasoning because an unanswered question is not evidence for God.
6. Even if the origin of life remained unsolved, that would not prove the personal God of religion.
7. Honest uncertainty is better than premature supernatural certainty.

## The Case So Far

The previous chapter showed that the human body is best understood through evolution, compromise, and biology, not perfect design. This chapter examined the earlier question of how life began and showed that an unfinished scientific problem should not be turned into evidence for God.

## Key Takeaway

The origin of life is a serious scientific question, but it is not proof of God. Science does not yet know every step from chemistry to biology, but ignorance is not evidence. A gap in knowledge should lead to investigation, not supernatural certainty. Life can be studied as a natural phenomenon without invoking divine creation.

## Sources and Further Reading


1. Robert M. Hazen, **Genesis: The Scientific Quest for Life's Origin**.
2. Nick Lane, **The Vital Question**.
3. David Deamer, **First Life**.
4. John Maynard Smith and Eörs Szathmáry, **The Major Transitions in Evolution**.
5. Jack W. Szostak, writings on protocells and early life.
6. Gerald F. Joyce, writings on RNA and early life.
7. Carl Sagan, **The Demon-Haunted World: Science as a Candle in the Dark**.


## Transition to the Next Chapter

The origin of life raises a difficult scientific question, but suffering raises a deeper moral one. If a loving and all-powerful God governs the world, why does innocent suffering occur so blindly? The next chapter turns from chemistry to pain, and from explanation to compassion.

# Chapter 9: The Unanswered Cry of Suffering

## The Question No Theology Can Escape

A child lies in a hospital bed.

The room is quiet except for the soft sounds of machines, footsteps, and whispered family prayers. The parents are exhausted. They have done everything they can. Doctors have explained the diagnosis. Nurses have given medications. Tests have been repeated. Treatments have been tried. The family has prayed with all the sincerity a human heart can hold.

Still, the disease continues.

At that bedside, the question of God stops being abstract.

It is no longer only about philosophy, scripture, or the origin of the universe. It becomes a direct moral question:

If a loving, all-powerful God exists, why does innocent suffering continue?

This question cannot be dismissed as lack of faith. It cannot be answered by quoting a verse and moving on. It cannot be softened by saying that humans are too small to understand. The suffering is real. The fear is real. The child's pain is real. The parents' grief is real.

Any belief in a personal God must face this reality honestly.

If God is only a vague name for mystery, the problem is different. But religion usually claims far more. It claims that God is loving, just, powerful, wise, and involved in human life. It claims that God hears prayer, cares for the weak, protects the innocent, and gives meaning to suffering.

The world does not clearly show this.

The world shows something far more painful.

Suffering falls without moral fairness.

## Why Innocent Suffering Matters

Not all suffering creates the same challenge for belief in God.

If a person harms others and later faces consequences, people may see some moral order. If a drunk driver causes an accident and is punished, we understand responsibility. If cruelty leads to social rejection, we understand cause and effect. Human choices can create suffering, and human beings can be morally responsible for their actions.

But innocent suffering is different.

A newborn with a genetic disease has not chosen wrongdoing.

A child with leukemia has not committed a sin deserving illness.

A family crushed by an earthquake has not been selected by moral guilt.

An animal torn apart in nature has not rebelled against God.

A person born into poverty has not earned hunger by spiritual failure.

This kind of suffering creates a deep wound in religious belief.

If God is all-knowing, he sees it.

If God is all-powerful, he could prevent it.

If God is perfectly loving, he should want to prevent it.

Yet the suffering continues.

The problem is not that life contains some difficulty. Growth, effort, learning, and responsibility may involve struggle. The deeper problem is extreme, undeserved suffering: pain that breaks bodies, destroys families, terrifies children, and serves no clear moral purpose.

Such suffering fits a natural world.

It does not fit easily with a loving ruler of the universe.

## Disease Does Not Follow Sin

In medicine, one lesson becomes clear again and again:

Disease follows biology, not moral guilt.

Cancer does not check whether a child has sinned.

Bacteria do not ask whether a patient is virtuous.

Viruses do not avoid kind people.

Genetic mutations do not wait for wrongdoing.

Autoimmune disease does not measure spiritual purity.

Stroke, dementia, heart failure, kidney failure, and respiratory failure do not follow a moral scorecard.

Disease has causes, but they are natural causes: genes, cells, infections, inflammation, aging, environment, chance, behavior, poverty, access to care, and biology. Some illnesses are influenced by human choices. Many are not. Even when behavior contributes, disease is not divine judgment. A smoker with lung cancer still deserves compassion. A person with diabetes still deserves care. A person with addiction still deserves treatment, not contempt.

Religious explanations often blur this truth.

They may speak of sin, karma, divine will, spiritual lessons, curses, or punishment. Such explanations can sound meaningful from a distance. But at the bedside, they can become cruel.

The sick do not need blame added to pain.

They need care.

A natural understanding of disease is not cold. It is kinder because it removes imaginary guilt. It allows us to say, “This happened because bodies are fragile, not because you deserved it.”

That is a deeply humane truth.

## The Cruelty of Blaming the Victim

When religion tries to protect God from the reality of suffering, it often shifts the burden onto human beings.

The sick are told they lacked faith.

The poor are told they are paying for karma.

The grieving are told God needed their loved one.

The disabled are told there is a hidden spiritual lesson.

The disaster victim is told the event was judgment.

The depressed person is told to pray harder.

The abused person is told to forgive too quickly or endure for spiritual reasons.

These responses may be intended to comfort, but they often harm.

They make suffering heavier by adding guilt, shame, or confusion. They ask the sufferer not only to endure pain, but also to defend the goodness of God. They turn compassion away from the human being and toward theology.

This is morally backwards.

The first duty in the presence of suffering is not to protect a belief.

It is to protect the person.

When a child is ill, the central question should not be, "How can this fit God's plan?" The central question should be, "How can we reduce this suffering?"

When people are hungry, the question should not be whether they deserve their condition. It should be how to feed them, protect them, educate them, and build a fairer society.

When victims are harmed by disaster, the question should not be whether heaven is angry. It should be how to rescue, rebuild, prepare, and prevent greater harm next time.

A moral response begins with the sufferer.

Not with defending God.

## Natural Disasters and Moral Blindness

Earthquakes do not strike by moral worth.

Floods do not spare children.

Hurricanes do not avoid hospitals.

Wildfires do not check prayer records.

Tsunamis do not separate believers from unbelievers.

Natural disasters reveal the moral blindness of nature.

They follow geology, weather, climate, ocean movement, gravity, heat, pressure, and physical law. These forces are not evil. They have no intention. But they are also not loving. They do not protect innocence. They do not reward virtue. They do not punish guilt.

This is difficult for belief in a personal God.

If disasters are purely natural, they fit a natural world. But if God governs the world with justice and love, their blind destruction becomes hard to explain.

Some believers say disasters are warnings or punishments. But this answer quickly becomes morally ugly. It implies that victims somehow deserved what happened, or that God was willing to kill children and innocent people to send a message.

Others say disasters are part of a larger plan. But if any human ruler allowed such destruction while having the power to prevent it, we would not call that ruler loving.

We would ask why he did nothing.

The same moral question must be asked of God.

## Animal Suffering Before Human Beings

The problem of suffering does not begin with human history.

Animals suffered for hundreds of millions of years before human beings existed.

They were hunted, wounded, infected, starved, frozen, burned, drowned, crushed, and eaten alive. Parasites invaded bodies. Teeth tore flesh. Bones broke. Young animals died. Disease spread. Fear and pain existed long before any human prayer, sin, scripture, or moral choice.

This matters because many religious explanations for suffering depend on human wrongdoing.

But animal suffering came first.

A fawn dying in a forest fire did not choose evil.

A fish eaten alive did not disobey God.

A dinosaur with infection did not need soul-making.

A parasite blinding an animal did not serve human free will.

Evolution explains this suffering clearly. Life survives through competition, reproduction, predation, adaptation, and death. Nature is not designed for universal happiness. It is shaped by survival.

But theism must explain why a loving creator would build a world where suffering was woven into life long before human beings appeared.

That is not easy.

Animal suffering shows that pain is not only a result of human moral failure. It is built into the natural history of life.

## "God Has a Plan" Is Not an Answer

When suffering becomes unbearable, people often say, "God has a plan."

The phrase is meant to comfort.

Sometimes it does. It may help people endure what they cannot change. It may give structure to grief. It may make tragedy feel less random.

But as an explanation, it fails.

If a plan cannot be examined, questioned, understood, or morally evaluated, then it explains nothing. It simply asks the sufferer to trust what appears cruel, pointless, or indifferent.

Would we accept this reasoning anywhere else?

If a doctor knowingly allowed a child to suffer when he could easily stop it, we would not excuse him by saying, “He must have a plan.”

If a parent watched a child drown and did nothing, we would not call the parent loving because of a hidden purpose.

If a ruler allowed preventable suffering on a massive scale, we would not praise his wisdom because his reasons were secret.

Yet religion often asks us to say this about God.

The hidden-plan answer protects belief, but it does not answer suffering.

It may silence the question.

It does not solve it.

## Free Will Cannot Carry the Burden

Free will is often offered as the answer to suffering.

Human beings are free, the argument says, and freedom allows evil. God could prevent all harm only by turning people into machines.

This may explain some human cruelty.

It does not explain most natural suffering.

Free will does not explain childhood cancer.

It does not explain genetic disease.

It does not explain earthquakes.

It does not explain tsunamis.

It does not explain animal suffering before humans.

It does not explain viruses, parasites, birth defects, or dementia.

Cancer cells have no free will. Earthquakes make no moral choices. Viruses do not rebel against God. Genetic mutations do not sin.

Free will cannot carry the burden of natural evil.

This deserves its own careful discussion, and the next chapter will examine it more fully. Here, the main point is simple: much suffering happens without any meaningful human choice behind it.

A world containing freedom did not have to contain so much blind natural pain.

## Prayer and the Silence Around Pain

In suffering, people pray.

This is deeply human. Prayer may express fear, love, helplessness, hope, and grief. No one should mock a frightened parent praying beside a sick child.

But prayer creates another problem for belief in God.

If God hears and answers prayer, why do the outcomes look so much like medicine, biology, chance, and circumstance?

Some people pray and recover.

Some people pray and die.

Some people are prayed for by thousands and still suffer.

Some people never pray and recover.

If God were answering prayers in a clear and reliable way, we should see stronger patterns. Instead, prayer often appears emotionally meaningful but not medically reliable.

Believers may say God answers in different ways: yes, no, or wait. But if every outcome can be called an answer, then prayer cannot be tested. It becomes impossible to distinguish divine response from no response.

Prayer will be discussed more fully later. But suffering raises the issue immediately.

At the bedside, we must ask whether God is truly acting, or whether human beings are speaking into silence while medicine and nature decide the outcome.

## The Emotional Honesty of Atheism

Atheism cannot remove suffering.

It cannot bring back the dead. It cannot promise heaven. It cannot say that every tear will be wiped away by a divine hand. It cannot make tragedy part of a perfect plan.

This may feel hard.

But atheism has one moral strength: it does not need to explain suffering as good.

It can say:

This is terrible.

This is unfair.

This person did not deserve it.

This child is not being punished.

This disease is not sacred.

This disaster is not justice.

This pain should be reduced.

There is honesty in that.

Atheism does not ask the sufferer to defend God. It does not ask a grieving parent to call loss a blessing. It does not tell the poor that hunger is their karma. It does not tell the sick that illness is divine punishment.

It allows compassion to stand without theological confusion.

Sometimes the most moral answer is not an explanation.

Sometimes it is presence, care, medicine, justice, and love.

## Human Beings Are the Help We Have

When suffering occurs, help usually comes from human beings.

Doctors diagnose. Nurses care. Researchers study. Families comfort. Strangers donate blood. Firefighters rescue. Engineers design safer buildings. Public health workers prevent disease. Volunteers bring food. Courts seek justice. Teachers educate. Scientists develop vaccines. Communities rebuild.

These efforts are imperfect because human beings are imperfect.

But they are real.

If a child is sick, we do not rely only on prayer. We seek doctors, medicines, hospitals, laboratories, imaging, surgery, chemotherapy, antibiotics, oxygen, fluids, and careful monitoring.

If a building collapses, we do not only ask why God allowed it. We search rubble, treat injuries, study construction, improve codes, and prepare better.

If disease spreads, we do not only perform rituals. We study transmission, sanitation, vaccines, behavior, treatment, and prevention.

This is how people actually live.

Even many believers act as if human help is necessary because divine help is not reliable enough to replace it.

That fact is important.

Human beings build systems of care because nature does not protect us automatically, and God does not visibly intervene in a dependable way.

Our responsibility is not a small matter.

It is the only help we can verify.

## Why Suffering Supports Atheism

The problem of suffering does not prove with mathematical certainty that no God of any kind exists.

But it strongly challenges the personal God of religion.

If God were loving, powerful, wise, and involved, the world should not contain so much blind, undeserved suffering. Disease should not strike children without cause. Disasters should not destroy without moral pattern. Animals should not suffer for millions of years before humans. Prayer should not look so much like chance. The innocent should not appear so unprotected.

A natural world explains these facts more clearly.

Biology produces disease.
Geology produces earthquakes.
Weather produces storms.
Evolution produces predation and extinction.
Genes mutate.
Cells fail.
Bodies age.
Chance matters.
Human choices matter.
Human systems matter.

No hidden moral plan is required.

A world without a personal God would look exactly like this: beautiful in places, terrible in places, lawful but indifferent, capable of life and suffering, offering no guaranteed justice except what conscious beings create for one another.

That is the world we see.

## Compassion Without Illusion

Letting go of the God explanation does not make us less compassionate.

It can make compassion clearer.

We no longer need to ask whether suffering was deserved. We no longer need to defend a divine plan. We no longer need to explain disease as punishment or disaster as warning. We no longer need to call cruelty mysterious goodness.

We can focus on the sufferer.

The child in pain needs relief.

The hungry need food.

The grieving need presence.

The sick need treatment.

The oppressed need justice.

The lonely need companionship.

The dying need comfort.

That is enough.

Compassion does not need a supernatural foundation. It begins with the simple recognition that another being can suffer and that suffering matters.

The absence of God does not make suffering meaningless.

It makes our response more urgent.

## Chapter Summary

1. Innocent suffering is one of the deepest challenges to belief in a loving, all-powerful God.
2. Disease follows biology, not sin, karma, curse, or divine punishment.
3. Religious explanations can harm when they blame victims or defend suffering as part of a hidden plan.
4. Natural disasters and animal suffering show that pain often occurs without moral purpose.
5. Free will cannot explain much natural suffering.
6. Prayer may comfort, but suffering does not show reliable divine rescue.
7. Atheism allows a morally honest response: suffering should not be defended; it should be reduced.

## The Case So Far

The previous chapter showed that the origin of life remains a scientific question, not evidence for God. This chapter turned from scientific mystery to moral reality. It showed that innocent suffering fits a natural world far better than a world governed by a loving, all-powerful personal God.

## Key Takeaway

The suffering of the innocent is not a minor problem for religion. It is a central moral challenge. Disease, disaster, genetic illness, animal pain, and failed prayer do not behave like divine justice. They behave like nature. A humane response does not defend suffering as God's plan. It works to reduce suffering wherever possible.

## Sources and Further Reading

1. David Hume, **Dialogues Concerning Natural Religion**.
2. J. L. Mackie, **The Miracle of Theism**.
3. William L. Rowe, "The Problem of Evil and Some Varieties of Atheism."
4. Paul Draper, writings on pain, pleasure, and the hypothesis of indifference.
5. Bart D. Ehrman, **God's Problem**.
6. J. L. Schellenberg, **Divine Hiddenness and Human Reason**.
7. Randolph M. Nesse and George C. Williams, **Why We Get Sick**.

## Transition to the Next Chapter

Many believers answer suffering by appealing to free will. They say God allows evil because human freedom is necessary. That answer may explain some human cruelty, but can it explain cancer, earthquakes, genetic disease, viruses, and animal suffering before human beings existed? The next chapter examines why free will cannot explain natural evil.

# Chapter 10: Why Free Will Cannot Explain Natural Evil

## When No Human Choice Is Involved

A storm gathers over the ocean. Warm water feeds it. Winds organize. Pressure falls. Clouds rise. The storm becomes a hurricane. It moves toward land.

No one has chosen evil.

No one has sinned in order to create the storm.

No demon is required.

No human will is guiding the wind.

Yet homes may be destroyed. Families may be displaced. Hospitals may lose power. Children may drown. The elderly may die alone. A whole community may be changed in one night.

This is natural suffering.

It is not caused by human cruelty. It is not the result of murder, theft, hatred, war, abuse, or selfishness. It comes from the ordinary forces of nature: heat, pressure, water, wind, gravity, geology, biology, and chance.

This is where one of religion's most common answers begins to fail.

Many believers say God allows evil because human beings have free will. Without freedom, they argue, human beings would be like machines. Love, morality, courage, and responsibility would be impossible. If people are free, they can choose goodness, but they can also choose cruelty.

This answer may explain some human evil.

It does not explain natural evil.

Free will cannot explain cancer cells, viruses, earthquakes, birth defects, floods, genetic disease, parasites, dementia, or animal suffering before human beings existed.

When no human choice is involved, free will loses its power as an answer.

## Moral Evil and Natural Evil

To understand the issue, we must separate two kinds of suffering.

The first is moral evil. This is suffering caused by human choices: murder, rape, abuse, torture, corruption, cruelty, war, oppression, neglect, exploitation, and betrayal.

The second is natural evil. This is suffering caused by natural processes: disease, earthquakes, tsunamis, hurricanes, drought, genetic disorders, animal predation, birth defects, pandemics, aging, and death.

The free will defense is aimed mostly at moral evil.

If a person freely chooses cruelty, the blame belongs to that person, not directly to God. This does not solve every problem, but at least the argument has some connection to human responsibility.

Natural evil is different.

A newborn with a metabolic disease did not choose it.

A child with bone cancer did not misuse freedom.

A patient with dementia did not freely decide to lose memory.

A village destroyed by earthquake did not cause tectonic plates to move.

An animal dying slowly from infection did not rebel against heaven.

These events are not moral choices.

They are natural processes.

Any defense of God must explain not only why human beings can harm one another, but why nature itself produces so much pain.

Free will does not answer that.

## Cancer Has No Free Will

Cancer is a disease of cells.

Cells divide. DNA is copied. Mistakes occur. Repair systems work most of the time, but not always. Some mutations affect growth control. A cell may begin to multiply when it should stop. It may invade nearby tissue. It may spread to distant organs. It may resist the body's normal signals.

This is not moral rebellion.

It is biology.

Cancer cells do not choose evil. They do not hate the patient. They are not punished sinners. They are living cells whose control systems have failed.

This matters because cancer often strikes the innocent. Children can develop leukemia. Young adults can develop sarcoma. A non-smoker can develop lung cancer. A careful person can develop lymphoma. Families can do everything right and still face devastating disease.

Free will explains none of this.

A loving and all-powerful God could allow human freedom without allowing children to suffer from cancer.

Human moral responsibility does not require uncontrolled cell division. Compassion, courage, and love do not require tumors in the bones, brain, blood, or lungs.

Cancer fits a natural body shaped by evolution, mutation, aging, and cellular vulnerability.

It does not fit easily into the free will defense.

## Viruses, Bacteria, and Blind Biology

Infectious diseases also expose the weakness of the free will answer.

Viruses enter cells and use their machinery to replicate. Bacteria grow, spread, invade, and sometimes produce toxins. Parasites live at the expense of their hosts. Mosquitoes transmit disease. Contaminated water spreads infection. Crowding, poverty, climate, sanitation, and biology shape epidemics.

None of this requires human evil.

Some disease spreads because of human choices, such as poor sanitation, war, neglect, misinformation, or lack of public health. But the existence of pathogens themselves is not explained by free will.

A virus does not need moral freedom in order to infect a child.

A bacterium does not need sin in order to cause sepsis.

A parasite does not need rebellion in order to blind or weaken a host.

If God designed or governs nature, then the structure of infectious disease becomes a serious question. Why would a loving creator build a world where life feeds on life at the microscopic level, where children can die from organisms too small to see, and where survival often depends on one organism exploiting another?

Evolution explains this clearly.

Organisms reproduce, compete, adapt, and exploit opportunities. Pathogens survive by using hosts. This is not moral. It is biological.

Free will is not involved.

## Birth Defects and Genetic Disease

Some suffering begins before birth.

A chromosome may be missing, extra, or rearranged. A single gene may carry a harmful mutation. A developmental process may go wrong. An organ may form incorrectly. A baby may be born with a damaged heart, malformed spine, impaired metabolism, or severe neurological disease.

Where is free will here?

The child has not chosen.

The parents may have done nothing wrong.

The gene has no moral intention.

The developing embryo is not making ethical decisions.

This kind of suffering is especially difficult for religious explanations because it begins before personal responsibility. A newborn does not deserve punishment. A fetus has not misused freedom. A genetic mutation is not a moral act.

Some traditions try to explain such suffering through past-life karma, inherited sin, or mysterious divine purpose. But these explanations often protect theology by placing hidden guilt on the innocent.

A more honest answer is biology.

Genes copy imperfectly. Development is complex. Bodies form through delicate processes. Sometimes those processes fail. The result can be tragic, but it is not morally deserved.

Free will does not explain birth defects.

Compassion should replace blame.

## Natural Disasters Do Not Protect Freedom

Some believers argue that God must allow stable natural laws for human freedom to exist. If God constantly interrupted nature, they say, human beings could not make meaningful choices. Fire must burn. Gravity must pull. Water must flow. The world must be predictable.

There is some truth here.

A stable world is necessary for action, learning, science, responsibility, and ordinary life. If gravity worked one day and stopped the next, human beings could not function. If fire sometimes burned and sometimes did not, we could not reliably act.

But this does not explain the scale of natural suffering.

A stable world does not require earthquakes that crush children under buildings.

It does not require tsunamis that drown thousands.

It does not require hurricanes that destroy hospitals.

It does not require volcanic eruptions that bury communities.

It does not require droughts that bring famine.

A world could be orderly and still contain far less suffering.

God, if all-powerful, would not have to abolish natural law in order to reduce natural disaster. A slight change in geology, weather patterns, disease mechanisms, or human vulnerability could prevent immense pain while leaving freedom intact.

Human freedom does not depend on the exact amount of natural destruction we observe.

The appeal to stable laws explains why nature has regularity.

It does not explain why regularity includes so much blind suffering.

## Animal Suffering Before Humans

Free will also fails because suffering existed long before human beings.

For millions of years, animals experienced injury, hunger, fear, infection, predation, parasites, cold, heat, drought, and death. Some animals were eaten alive. Some died slowly from wounds. Some starved. Some were trapped. Some lost offspring. Some suffered disease.

No human choice caused this.

No human sin existed.

No human moral freedom was being preserved.

This is a major problem for the free will defense.

If animal suffering began before human beings, then it cannot be explained as the result of human misuse of freedom. It belongs to the structure of life itself.

Evolution explains why this happened.

Life reproduces, competes, adapts, eats, is eaten, survives, and dies. Nature is not arranged for universal comfort. It is shaped by survival and reproduction.

But why would a loving God create such a system?

Why use millions of years of pain, fear, predation, extinction, and struggle as the path to life?

Free will does not answer.

Animal suffering shows that natural evil is older than human morality.

## The "Soul-Making" Answer

Another religious answer says suffering helps build the soul.

Pain teaches courage. Struggle teaches patience. Loss teaches compassion. Difficulty develops maturity. Without suffering, human beings would remain shallow.

There is some truth here.

Human beings can grow through hardship. Many people become stronger, wiser, or more compassionate after difficulty. Some suffering can teach. Some struggle can deepen character.

But this answer has serious limits.

Not all suffering improves the sufferer.

Some suffering destroys.

A child who dies in agony does not grow into moral maturity. A patient with advanced dementia does not become spiritually stronger by losing memory and identity. A newborn with a fatal genetic disease does not need character development. An animal eaten alive is not being prepared for moral wisdom.

Even when suffering can lead to growth, that does not justify all suffering.

A teacher does not need to break a child's bones to teach courage. A parent does not need to allow preventable disease to teach love. A doctor does not create infection in order to teach patience.

If we would not defend such behavior in human beings, we should be careful before defending it in God.

The soul-making answer may explain some forms of ordinary struggle.

It cannot justify the vast amount of extreme, pointless, and innocent suffering in the natural world.

## Could God Intervene Without Destroying Freedom?

Religious defenses often assume that if God intervened to prevent suffering, freedom would disappear.

But this is not convincing.

Human beings already accept many limits on harm without believing freedom is destroyed.

We prevent children from drinking poison.

We rescue people from drowning.

We vaccinate against disease.

We design safer cars.

We build earthquake-resistant buildings.

We use antibiotics.

We remove tumors.

We treat pain.

We stop violence when we can.

None of this destroys freedom. It reduces suffering.

If human beings can reduce suffering without destroying moral responsibility, an all-powerful God could do far more.

God could prevent a child's leukemia without making human beings into machines.

God could stop a deadly mutation before birth without removing moral choice.

God could reduce the cruelty of animal predation without eliminating love, courage, or responsibility.

God could make the world safer while still allowing meaningful life.

The fact that we can imagine many improvements to the world is itself important.

If the world were designed by perfect wisdom and love, it should not be so easy to see ways it could contain less innocent suffering.

## The Problem of Divine Design

Natural evil becomes especially difficult when religion claims that God designed the world.

If God merely watches a natural world he did not create, the question is different. But most religions claim that God created or governs nature. That means the structure of nature matters.

A world of viruses, parasites, cancer, genetic disease, earthquakes, and predation cannot be blamed only on human freedom.

The system itself produces suffering.

If God designed the system, then God designed a system in which suffering is built into the machinery of life.

Evolution explains this without moral contradiction. Natural selection is not guided by kindness. It preserves reproductive success, not happiness. It builds bodies that work well enough, not bodies free of disease. It produces ecosystems where life often feeds on life.

But a loving divine designer creates a deeper problem.

Why choose this method?

Why create through pain, extinction, and struggle?

Why build bodies vulnerable to mutation, infection, cancer, and decay?

Why make consciousness capable of terror and grief, then place it in a world where suffering is everywhere?

Free will cannot answer these questions because the questions are about design, not choice.

## The Cost of Stretching Free Will Too Far

A good explanation should fit the problem it is trying to solve.

Free will may help explain why human beings can lie, steal, betray, abuse, or murder. It does not explain why nature produces suffering without moral choice.

When free will is stretched to explain everything, it becomes weak.

It becomes a blanket answer placed over very different problems. But cancer is not murder. An earthquake is not theft. A birth defect is not betrayal. A parasite is not cruelty in the moral sense. Dementia is not sin.

Different causes need different explanations.

Religion often uses free will because it protects God from responsibility for evil. But when the suffering is natural, the protection fails.

The world contains too much pain that has nothing to do with human freedom.

That pain must be faced honestly.

The free will defense does not do that.

It changes the subject from nature's suffering to human choice.

## Why Natural Evil Supports Atheism

Natural evil strongly supports science-based atheism because it fits a natural world so well.

In a natural world, bodies are vulnerable because they are biological.

Diseases arise because cells, genes, and organisms operate through natural processes.

Disasters occur because Earth is governed by geology, weather, heat, pressure, water, and gravity.

Animals suffer because evolution works through survival, reproduction, competition, and death.

Aging occurs because bodies are imperfect systems with limited repair.

There is no need to add sin, punishment, karma, demons, divine testing, or hidden plans.

The natural explanation is clearer.

It does not make suffering good. It does not make suffering easy. It simply explains why suffering occurs without pretending that it is morally deserved or divinely planned.

A world without a personal God would contain exactly this kind of suffering: lawful, blind, uneven, often cruel, and indifferent to innocence.

That is the world we observe.

## What This Means for Human Responsibility

If natural evil is not caused by divine justice or human sin, then our responsibility becomes more urgent.

We should not explain disasters as God's will.

We should build safer homes, better warning systems, stronger public health, and more compassionate rescue systems.

We should not explain disease as punishment.

We should support medicine, prevention, research, vaccination, screening, sanitation, nutrition, and fair access to care.

We should not explain mental illness as spiritual failure.

We should offer treatment, understanding, and dignity.

We should not explain poverty as karma.

We should study social conditions, education, inequality, opportunity, policy, and justice.

Natural explanation does not make us passive.

It calls us to action.

If no divine power is reliably preventing natural suffering, then human knowledge, cooperation, and compassion matter enormously.

We are not responsible for every disaster.

But we are responsible for how we respond.

# Chapter Summary

1. Free will may explain some human-caused evil, but it does not explain natural evil.
2. Cancer, viruses, genetic disease, birth defects, earthquakes, and animal suffering are not caused by moral choice.
3. Stable natural laws do not require the amount of innocent suffering we observe.
4. Animal suffering existed long before human beings, so it cannot be explained by human sin or freedom.
5. The soul-making answer fails when suffering destroys rather than teaches.
6. An all-powerful God could reduce natural suffering without destroying human freedom.
7. Natural evil fits a natural world better than a world governed by perfect love and power.

## The Case So Far

The previous chapter showed that innocent suffering is a central moral challenge to belief in God. This chapter examined one of the most common religious replies, the free will defense, and showed why it cannot explain natural suffering.

## Key Takeaway

Free will cannot explain natural evil. Cancer cells, viruses, earthquakes, genetic mutations, birth defects, parasites, and prehuman animal suffering do not make moral choices. They follow nature. A world filled with blind natural suffering is difficult to reconcile with a loving, all-powerful God, but it makes clear sense in a natural universe.

## Sources and Further Reading


1. David Hume, **Dialogues Concerning Natural Religion**.
2. J. L. Mackie, **The Miracle of Theism**.
3. William L. Rowe, "The Problem of Evil and Some Varieties of Atheism."
4. Paul Draper, writings on pain, pleasure, and the hypothesis of indifference.
5. Bart D. Ehrman, **God's Problem**.
6. Randolph M. Nesse and George C. Williams, **Why We Get Sick**.
7. Charles Darwin, **On the Origin of Species**.


## Transition to the Next Chapter

Natural evil weakens the free will defense. But another problem remains. If God exists and wants human beings to know him, why is he so hidden? Why do sincere seekers often meet silence? The next chapter examines the silence of God.

# Chapter 11: The Silence of God

## The Prayer That Receives No Answer

There are moments when a human being calls out with complete sincerity.

A parent prays for a child.
A patient prays before surgery.
A grieving person asks for one sign.
A lonely person begs not to feel abandoned.
A doubter asks honestly, "God, if you are real, help me know."

Sometimes the person receives comfort. Sometimes life improves. Sometimes a coincidence feels meaningful. Sometimes the heart becomes calm.

But often there is only silence.

No clear voice.
No unmistakable sign.

No answer that can be distinguished from emotion, chance, memory, or hope.

For many believers, this silence is painful. For many doubters, it is decisive.

If God is supposed to be personal, loving, and interested in relationship, why is he so hidden? Why does belief depend so much on birthplace, childhood teaching, interpretation, fear, and community? Why do sincere seekers often remain unconvinced?

This is the problem of divine hiddenness.

It is different from the problem of suffering, though the two are connected. Suffering asks why a loving God allows so much pain. Hiddenness asks why a loving God remains so unclear.

A God who wants to be known should not be so difficult to find.

## What Divine Hiddenness Means

Divine hiddenness does not mean that no one believes in God.

Billions do.

It does not mean that no one feels God.

Many people do.

It does not mean that religion has no emotional power.

It clearly does.

The problem is different.

The problem is that God is not clearly available as a public reality. Belief is uncertain, disputed, culturally shaped, and open to many interpretations. One person sees God in an event. Another sees coincidence. One person sees answered prayer. Another sees biology or chance. One person sees divine revelation in a scripture. Another sees a human text.

If God existed as the personal ruler described by religion, we might expect clearer evidence.

Not coercive evidence. Not overwhelming force that removes freedom. But enough clarity that sincere, honest people would not be left guessing whether anyone is there.

Instead, the world looks ambiguous.

God is believed, imagined, defended, interpreted, and felt.

But God is not plainly seen.

That is the difficulty.

## The Loving Parent Analogy

The hiddenness problem becomes clearer when we think of a loving parent.

A good parent does not hide from a child who is frightened, confused, or searching. A good parent may allow a child to grow, learn, and make choices, but does not disappear completely and then blame the child for not feeling loved.

If a child cries in the dark, a loving parent responds.

If a child is lost, a loving parent tries to be found.

If a child sincerely asks, "Are you there?" a loving parent does not remain silent for years and then punish the child for doubt.

Religious believers may say God is not like a human parent. That may be true in some ways. But religion often calls God father, protector, guide, shepherd, healer, and loving presence. If those words mean anything, hiddenness becomes a real problem.

Love normally seeks connection.

Love makes itself known.

Love does not require the beloved to solve a puzzle of geography, scripture, culture, and fear before relationship can begin.

A God who hides so thoroughly from sincere seekers does not resemble perfect love.

He resembles absence.

## Belief Depends Too Much on Geography

One of the strongest signs of hiddenness is the geography of belief.

A child born in Saudi Arabia is very likely to become Muslim.

A child born in India may become Hindu, Muslim, Sikh, Christian, Jain, Buddhist, or secular depending greatly on family and region.

A child born in rural Mississippi is more likely to become Christian.

A child born in Thailand may become Buddhist.

A child born in a secular Scandinavian family may grow up with little religious belief.

This pattern is not absolute. People convert. People doubt. People leave religion. People change.

But the broad pattern is clear.

Religious belief follows birth, family, culture, language, history, and community.

If one loving God had clearly revealed one essential truth to humanity, why would belief depend so strongly on where a child is born? Why would eternal truth be distributed like cultural inheritance?

This does not look like clear divine communication.

It looks like human transmission.

People usually inherit the religion of their surroundings, then later call it truth.

That is exactly what we would expect if religions were human-made.

It is harder to explain if one God urgently wants all people to know him clearly.

## The Problem of Conflicting Revelations

Hiddenness deepens when we look at revelation.

Religions do not merely differ in minor details. They disagree about God, salvation, worship, sin, scripture, prophets, rituals, afterlife, morality, and the nature of ultimate reality.

One religion says God became human.

Another says this is impossible.

One says there is one life followed by judgment.

Another says there is rebirth.

One says salvation comes through faith.

Another says liberation comes through knowledge, devotion, discipline, or awakening.

One says a scripture is final.

Another says a different scripture is final.

One says God is personal.

Another says ultimate reality is beyond personhood.

These contradictions matter.

If God has revealed truth, why does revelation look so confused? Why does the world contain many sincere communities, each convinced that its inherited revelation is true and others are mistaken?

A loving God could speak more clearly than this.

Human beings can write, translate, broadcast, teach, record, and clarify. We can send messages across continents and into space. We can preserve instructions, repeat warnings, and correct misunderstanding.

Surely an all-powerful God could communicate essential truth without leaving humanity divided into thousands of competing claims.

Conflicting revelations look less like divine clarity and more like human religion.

## Why Scripture Does Not End the Silence

Believers may answer, "God has spoken through scripture."

But scripture does not end hiddenness.

It often increases it.

Which scripture is truly divine?
How should it be interpreted?
Which passages are literal?
Which are symbolic?
Which commands are eternal?
Which belong to history?
Why do sacred texts contain moral problems?
Why do believers disagree so deeply about the same book?

A book does not prove itself by claiming authority. Every religious tradition can point to its own sacred texts. The existence of a scripture shows that human beings wrote, preserved, translated, and interpreted a text. It does not automatically prove that God spoke.

Scriptures also require human judgment. Believers must decide which passages to emphasize, soften, ignore, or reinterpret. They use moral reasoning, historical study, tradition, language, and community authority to make sense of the text.

That means the supposed divine message still comes through human filters.

If God wanted to speak clearly, scripture is a surprisingly unclear method. It leaves too much room for contradiction, conflict, violence, sectarian division, and honest confusion.

The silence remains.

It is covered with words, but not removed.

# "God Is Hidden to Preserve Free Will"

One common answer says God remains hidden to preserve human freedom.

If God were too obvious, people say, we would be forced to believe. Faith would no longer be free. Love would become obedience under pressure.

This answer sounds reasonable at first, but it fails on closer examination.

Knowledge does not destroy freedom.

A child can know his parent exists and still choose whether to obey, love, ignore, trust, or rebel. A citizen can know a law exists and still choose whether to follow it. A patient can know a doctor's advice and still decide whether to accept it.

Clear knowledge does not remove choice.

If God appeared clearly, people could still accept or reject his guidance. They could still love or not love. They could still obey or disobey. They could still act morally or selfishly.

In many religious stories, God is obvious to people, yet they still disobey. Angels, prophets, kings, and communities are said to know God and still resist. Even within religious imagination, knowledge of God does not eliminate freedom.

So why should ordinary sincere seekers be denied clarity?

Hiddenness does not protect free will.

It prevents informed choice.

## "God Wants Faith"

Another answer says God wants faith.

But why would a loving God value belief without sufficient evidence?

In ordinary life, we do not praise someone for believing serious claims without evidence. We do not call it a virtue when a patient trusts an untested treatment blindly. We do not admire a jury that accepts an accusation without proof. We do not teach children that all unsupported claims deserve belief.

Truth should not fear evidence.

If God is real, evidence should strengthen faith, not destroy it. A God who values honesty should welcome sincere inquiry. A God who gave human beings minds should not punish them for using those minds carefully.

Faith may have emotional value.

It may express loyalty, hope, or trust.

But if faith means believing without enough evidence, then it is a strange requirement for a just God to impose, especially when people are born into different religions and taught different sacred certainties.

A loving God should not make eternal truth depend on the accident of childhood belief.

## Honest Nonbelief

Many religious traditions treat nonbelief as rebellion.

The unbeliever is accused of pride, sin, hardness of heart, love of pleasure, hatred of God, or desire to avoid moral responsibility.

Sometimes this may be true of some individuals. Human motives are mixed.

But honest nonbelief exists.

There are people who sincerely seek truth and do not find God. They read, pray, think, study, doubt, ask, and struggle. They may want belief. They may fear losing belief. They may lose family peace, community, certainty, and comfort because they can no longer honestly accept what they were taught.

Such people are not necessarily wicked.

They may simply be unconvinced.

This matters.

If God is loving and just, honest nonbelief should not be treated as moral failure. A person should not be condemned for refusing to claim certainty where evidence is unclear.

If a belief is not supported strongly enough, withholding belief can be an act of honesty.

A God who punishes honest doubt would not be morally admirable.

And a God who allows honest seekers to remain in darkness is difficult to reconcile with perfect love.

## Hiddenness and Hell

The problem becomes even more serious when religion adds eternal punishment.

If disbelief can lead to hell, then divine hiddenness becomes morally unbearable.

A loving God who threatens severe punishment for unbelief should make the truth unmistakably clear. No just teacher gives a final exam while hiding the lesson. No fair judge punishes a person for failing to accept unclear evidence. No loving parent demands trust while refusing to answer the frightened child.

Some believers say hell is chosen by those who reject God.

But what about those who never clearly found God?

What about those born into other religions?

What about those taught frightening beliefs in childhood?

What about those who sincerely examined the evidence and remained unconvinced?

What about those harmed by religion and unable to see it as loving?

If the stakes are eternal, the evidence should be proportionate.

Hiddenness may be troubling in any form of theism.

Hiddenness combined with hell is morally much worse.

It asks human beings to accept that God hides clearly enough for honest doubt to exist, then punishes doubt as if clarity had been given.

That is not justice.

## The Silence in Suffering

Divine hiddenness is most painful in suffering.

It is one thing for God to be hidden in philosophical debate. It is another for God to be hidden beside a hospital bed, after a disaster, during abuse, in mental illness, or in grief.

People often do not ask for a full explanation.

They ask for help.

A child asks for pain to stop. A parent asks for one sign. A person trapped in despair asks not to feel alone. A victim asks for protection. A dying person asks for peace.

Sometimes human comfort comes. A nurse enters. A friend sits nearby. A doctor adjusts the medicine. A stranger donates. A family holds hands.

But God remains unclear.

Believers may interpret inner peace as God. Others may interpret it as the mind's response to love, exhaustion, medication, acceptance, or human presence. The evidence remains ambiguous.

This is the deep wound of hiddenness.

The world contains suffering intense enough to demand divine compassion, yet the divine presence remains indistinguishable from silence.

Human beings show up.

God does not clearly do so.

## Why Hiddenness Supports Atheism

Divine hiddenness supports atheism because it fits naturally with the absence of a personal God.

If there is no God, we would expect religions to arise through culture, fear, hope, family, and imagination.

We would expect belief to vary by geography.

We would expect scriptures to conflict.

We would expect prayer to feel meaningful to believers but produce no clear public pattern.

We would expect sincere seekers to disagree.

We would expect religious experiences to reflect upbringing and expectation.

We would expect God's silence because no divine person is speaking.

That is what we see.

Theism must explain why God hides, why revelation is confused, why sincere nonbelief exists, why geography predicts belief so strongly, why prayer is ambiguous, and why the world does not show a clear divine presence.

Atheism has a simpler explanation.

There is silence because there is no personal God answering.

That explanation may be emotionally difficult.

But it fits the evidence.

## Living Without a Hidden God

Letting go of a hidden God can be painful at first.

A person may feel loss, fear, loneliness, or guilt. He may miss prayer. He may miss the feeling of being watched over. He may grieve the idea that someone unseen was guiding every step.

That grief is real.

But something can also be gained.

A person no longer has to interpret silence as love. He no longer has to call confusion a test. He no longer has to blame himself for not hearing a voice that may not exist. He no longer has to pretend that unclear evidence is clear.

He can live honestly.

He can seek truth without fear.

He can find comfort in real relationships.

He can build meaning through love, work, service, learning, and compassion.

He can accept that human beings must be present for one another because no hidden rescuer can be relied upon.

This does not make life empty.

It makes human presence more important.

When God is silent, the answer is not to defend the silence.

The answer is to become less silent ourselves.

## Chapter Summary

1. Divine hiddenness asks why a loving personal God is so unclear to sincere seekers.

2. Belief in God depends strongly on geography, childhood, culture, and inherited religion.
3. Conflicting revelations and scriptures make divine communication appear human and uncertain.
4. Clear evidence of God would not destroy free will; it would make informed choice possible.
5. Honest nonbelief exists and should not be treated as moral failure.
6. Hiddenness becomes especially troubling when combined with eternal punishment.
7. The silence of God fits atheism more naturally than belief in a loving, personal, intervening God.

## The Case So Far

The previous chapter showed why free will cannot explain natural suffering. This chapter examined another major difficulty for belief: if God exists and wants to be known, his presence should be clearer than it is. Divine hiddenness makes the personal God of religion harder to defend.

## Key Takeaway

A loving God who wants relationship should not be so hidden that sincere seekers find only silence. Geography, conflicting revelations, unanswered prayer, honest nonbelief, and unclear evidence all point in the same direction. A hidden God and no God look too similar. The simpler explanation is that no personal God is speaking.

## Sources and Further Reading

1. J. L. Schellenberg, **Divine Hiddenness and Human Reason**.
2. Theodore M. Drange, **Nonbelief and Evil**.
3. David Hume, **Dialogues Concerning Natural Religion**.
4. J. L. Mackie, **The Miracle of Theism**.
5. Bertrand Russell, **Why I Am Not a Christian**.
6. Daniel C. Dennett, **Breaking the Spell: Religion as a Natural Phenomenon**.

# Transition to the Next Chapter

If God remains hidden, believers often turn to prayer and miracles as signs that he still acts in the world. But do prayer and miracle claims provide reliable evidence, or do they reflect hope, interpretation, coincidence, and selective memory? The next chapter examines that question.

# Chapter 12: Prayer, Miracles, and the Missing Evidence

## Where Prayer Feels Most Human

Prayer often begins where human strength ends.

A parent prays beside a sick child.
A patient prays before surgery.
A family prays after an accident.
A lonely person prays in the dark.
A grieving person prays for one more sign.
A frightened heart says, "Please help me."

No compassionate person should mock this.

Prayer can be one of the most human responses to fear, grief, helplessness, gratitude, and hope. It may calm the mind. It may help people endure pain. It may bring families together. It may give language to emotions that are too heavy to carry alone.

But the emotional value of prayer is not the same as evidence that prayer changes reality through divine power.

That distinction matters.

A prayer may comfort the person praying without proving that a God is listening. A ritual may steady the mind without proving supernatural intervention. A family may feel stronger after prayer without proving that the course of disease has changed.

The question in this chapter is not whether prayer feels meaningful.

For many people, it does.

The question is whether prayer and miracles provide reliable evidence for the personal God of religion.

## Comfort Is Not the Same as Evidence

Prayer can help people emotionally.

It may slow breathing, focus attention, reduce loneliness, create hope, and connect a person with family or community. In this way, prayer may function like meditation, reflection, music, silence, or conversation.

These effects can be real.

But they do not prove God.

A Buddhist may feel peace during meditation without belief in a creator God. A Hindu may feel blessed by a deity. A Christian may feel close to Jesus. A Muslim may feel comfort through Allah. A secular person may feel calm while sitting quietly, walking in nature, or speaking with a trusted friend.

The feeling of peace is real.

The interpretation differs.

If prayer brings calm, that tells us something about the human mind. It does not automatically tell us that a divine being has acted outside the mind.

This point should be made gently but clearly.

We can respect prayer as human expression while still asking whether it produces evidence of supernatural intervention.

## If Prayer Worked Reliably, Patterns Should Appear

If prayer affected events through divine power, we should expect some clear pattern.

Not every prayer would need to be answered exactly as requested. But if a loving God truly responded to prayer, especially sincere prayer for urgent need, the world should show some difference.

Patients prayed for should recover at rates that cannot be explained by medicine, biology, chance, and social support.

Natural disasters should spare prayerful communities in a way that can be detected.

Children prayed for by thousands should show outcomes beyond what medical science predicts.

Danger should be avoided more often when believers pray for protection.

But this is not what we see.

Believers recover, and believers die.

Atheists recover, and atheists die.

Prayers are offered before surgeries that succeed and before surgeries that fail.

Religious people suffer accidents, cancers, infections, miscarriages, strokes, and disasters like everyone else.

The world does not show a dependable prayer pattern.

This does not prove that no one ever feels helped by prayer. It means prayer does not appear to operate as a reliable force changing external reality.

If God answers prayer, the evidence is strangely hard to distinguish from natural outcome.

## The Problem With "Yes, No, or Wait"

Many believers respond by saying God answers every prayer in one of three ways:

Yes, no, or wait.

This may bring comfort, but it creates a serious problem.

If recovery occurs, prayer was answered yes.

If recovery does not occur, prayer was answered no.

If nothing happens yet, prayer was answered wait.

Under this system, every possible outcome becomes an answer. Success confirms prayer. Failure confirms prayer. Delay confirms prayer. Silence confirms prayer.

That means prayer cannot be tested.

A claim that explains every outcome explains very little.

Imagine if a medicine were defended this way. If the patient improves, the medicine worked. If the patient worsens, the medicine worked in a mysterious way. If nothing changes, the medicine is waiting. No responsible physician would accept that standard.

Prayer is often protected from failure by interpretation.

But if no possible result counts against the claim, then the claim is not evidence-based.

It is belief protected from examination.

## Prayer and Medicine

Prayer is often most intense during illness.

This is understandable. Illness makes human beings feel vulnerable. Patients and families want hope. They want strength. They want healing. They want someone beyond the hospital room to care.

But when illness improves, we should ask carefully what caused the improvement.

Was there surgery?
Was there chemotherapy?
Was there an antibiotic?
Was there oxygen, insulin, fluids, blood, dialysis, or intensive care?
Was the diagnosis uncertain?
Was the disease naturally variable?
Was the body's immune system involved?
Was the outcome within the range medicine would expect?

In clinical life, recovery often has many causes. Modern medicine, nursing care, early diagnosis, nutrition, sanitation, vaccines, technology, and the body's own biology can all affect outcome.

When a patient recovers after prayer and treatment, it is often God who receives the credit while the human work becomes invisible.

This is unfair to reality.

It ignores the scientists who discovered the treatment, the physicians who made decisions, the nurses who watched carefully, the technicians who ran tests, the donors who gave blood, the pharmacists who prepared medication, and the family members who provided care.

Gratitude should not erase human effort.

If medicine does the visible work, medicine deserves visible credit.

## The Study of Intercessory Prayer

Some researchers have tried to study whether prayer improves medical outcomes when people pray for patients from a distance.

The results have not provided strong evidence that intercessory prayer produces clear medical benefit beyond ordinary factors.

This should not surprise us.

If prayer worked as many believers imagine, it should not be difficult to detect. Large numbers of sincere prayers should produce measurable differences, especially in serious illness. Yet careful studies have not shown the kind of reliable effect that would be expected from divine intervention.

Believers may reply that God cannot be tested.

But this creates another problem.

If prayer changes outcomes, then outcomes can be examined.

If prayer does not produce observable differences, then it should not be claimed as evidence.

Religion cannot have it both ways. It cannot claim that prayer heals and then say healing cannot be tested when evidence is weak.

A God who truly intervenes in the physical world should leave more than emotional impressions.

## Miracle Stories

Miracle stories exist in every religion.

A tumor disappears. A person survives an accident. A child recovers unexpectedly. A statue seems to weep. A saint heals. A prayer brings money at the right time. A dream predicts an event. A person narrowly escapes danger. A disease improves after visiting a shrine.

Such stories are powerful.

They travel quickly because human beings love hope. They are repeated in families, temples, churches, mosques, shrines, books, and social media. They become part of religious identity.

But a story is not the same as evidence.

Many miracle stories rely on incomplete information. The diagnosis may have been uncertain. The disease may have a variable course. Other treatments may have been used. The memory may have changed. The rare recovery may be real but natural. Coincidences may be remembered when they seem meaningful and forgotten when they do not.

This does not mean every person who reports a miracle is lying.

Most are not.

It means testimony is not enough for extraordinary claims.

A miracle must be examined more carefully than an ordinary event because the claim is so large.

## Every Religion Has Miracles

Another problem is that miracle claims are not limited to one religion.

Christians report miracles.

Hindus report miracles.

Muslims report miracles.

Sikhs report miracles.

Buddhists report miracles.

Catholics, Protestants, Orthodox believers, gurus, saints, shamans, spiritual healers, and many religious movements all have miracle stories.

These claims cannot all prove their religions true, because the religions often contradict one another.

If miracle stories prove Christianity, what do Hindu miracles prove?

If Muslim miracle claims prove Islam, what do Catholic miracle claims prove?

If a healing at one shrine proves one deity, what does a healing at another shrine prove?

The more widely miracle stories appear, the less they support one specific religion.

They show something about human hope, interpretation, memory, and storytelling.

They do not clearly reveal one God.

A reliable divine message should not look so similar across conflicting traditions.

## Rare Events Are Not Automatically Miracles

Rare events happen.

Some cancers shrink unexpectedly. Some patients survive against the odds. Some people escape accidents by seconds. Some dreams seem to match later events. Some coincidences feel too meaningful to ignore.

But rare does not mean supernatural.

In a world with billions of people and countless events every day, unlikely things will happen regularly. Some will feel personal. Some will occur after prayer. Some will occur after rituals. Some will occur after no prayer at all.

Human beings remember the hits and forget the misses.

A person prays and receives help on one day, and the story is remembered for years. Many other prayers that lead to no result are quietly absorbed into silence, mystery, or "God's plan."

This selective memory makes miracles seem more common than they are.

A rare recovery may be medically unusual and still natural.

An unlikely escape may be fortunate and still not divine.

A coincidence may be emotionally meaningful and still not supernatural.

Before calling an event a miracle, we must ask whether natural explanations have truly been ruled out.

Usually, they have not.

## Why Are the Miracles So Small?

If an all-powerful God performs miracles, we should ask why the evidence is so limited.

Why do miracles so often appear at the edge of uncertainty: a recovery that might have happened naturally, a vague sign, a private vision, a coincidence, a statue, a dream, a story from long ago?

Why not clear miracles that remove doubt?

Why not regrow an amputated limb in public?

Why not restore a destroyed spinal cord instantly?

Why not end a plague overnight after prayer?

Why not stop a tsunami before it reaches children?

Why not reveal the same truth unmistakably to all humanity?

The problem is not that people ask for entertainment.

The problem is that religious claims are enormous, but the evidence is often ambiguous.

A God powerful enough to create galaxies should not be limited to events that can be explained by coincidence, psychology, medical uncertainty, or storytelling.

If miracles are meant to reveal God, they are strangely unclear.

If they are not meant to reveal God, they cannot be used as strong evidence.

## God Gets Credit, Humans Disappear

When something good happens, God often receives credit.

A surgery succeeds, and God is praised.

A patient survives sepsis, and it is called a miracle.

A medication works, and people thank heaven.

A vaccine prevents disease, and divine protection is invoked.

A rescue team saves victims, and God is said to have saved them.

Gratitude is understandable. But the habit can erase human effort.

It can make doctors, nurses, researchers, engineers, firefighters, public health workers, donors, volunteers, and families disappear behind divine language.

This matters morally.

If a child survives because of chemotherapy, antibiotics, blood transfusion, surgery, oxygen, and expert care, it is not disrespectful to thank the people and systems that made survival possible. It is truthful.

Human work does not become less meaningful because someone calls the result a miracle.

A science-based view gives credit where evidence points.

It honors the visible chain of care.

## The Harm of Failed Prayer

Prayer does not only raise intellectual questions.

It can also create emotional harm.

When prayer fails, people may blame themselves.

Did I not pray enough?
Was my faith too weak?
Was God displeased?
Did I have hidden sin?
Did I ask wrongly?
Did I fail my loved one?

This can be cruel.

A parent who loses a child should not carry the added burden of wondering whether stronger faith might have saved the child. A patient whose disease progresses should not feel spiritually responsible for biology. A grieving family should not be made to feel that God denied them because they prayed incorrectly.

Failed prayer can create guilt where there should be tenderness.

This is one reason prayer claims must be examined carefully.

It is not harmless to tell people that prayer changes outcomes when the evidence does not support that belief. Hope can strengthen people, but false hope can wound them.

Compassion should not require theological blame.

## Prayer Can Replace Action

Prayer can also become harmful when it replaces action.

If prayer comforts someone while medicine, justice, or practical help continues, it may cause little harm.

But if prayer substitutes for treatment, prevention, rescue, counseling, vaccination, public health, or protection from abuse, it can become dangerous.

A sick child needs medical care.

A victim of violence needs safety.

A person with severe depression may need professional help.

A community facing disease needs sanitation, science, and treatment.

A society facing injustice needs law, courage, education, and reform.

Prayer alone is not enough.

Even many believers understand this in practice. They pray, but they also go to hospitals. They pray, but they also call emergency services. They pray, but they also use antibiotics, surgery, insulin, vaccines, and blood pressure medicines.

This shows something important.

People may speak as if prayer is powerful, but they live as if human action is necessary.

They are right.

## The Psychological Power of Prayer

A natural explanation of prayer does not mean prayer has no effect at all.

Prayer can affect the person praying.

It can focus attention. It can slow the mind. It can provide a sense of release. It can help organize grief. It can connect someone to community. It can create hope during helplessness. It can become a form of meditation, confession, gratitude, or self-reflection.

These effects are real.

But they are psychological and social effects.

They do not prove divine intervention.

A person can experience similar benefits through meditation, breathing exercises, therapy, journaling, music, nature, community, or quiet reflection.

Understanding prayer naturally does not insult those who pray. It simply separates inner experience from external evidence.

Prayer may change the person.

That is different from proving that prayer changes the world through God.

## What Would a Real Miracle Show?

A genuine miracle, if offered as evidence for God, should be clear.

It should be public enough to examine.

It should not depend only on rumor.

It should not require ignoring natural explanations.

It should not appear only in one religious community.

It should be strong enough to distinguish divine action from coincidence, psychology, medical uncertainty, or selective memory.

For example, if amputated limbs regrew after prayer in documented cases, if blind eyes with destroyed anatomy were restored instantly, if fatal genetic diseases vanished in ways no biology could explain, if the same divine message appeared clearly to all humanity, then the discussion would change.

But that is not the evidence we have.

Instead, miracle claims are usually ambiguous, private, disputed, poorly documented, or explainable in more ordinary ways.

An all-powerful God could provide better evidence.

The absence of such evidence matters.

## A More Honest Response to Suffering

When people suffer, the most honest response is not to promise miracles.

It is to be present and to help.

Bring food.

Give blood.

Support treatment.

Sit beside the grieving.

Protect the abused.

Fund research.

Teach prevention.

Build safer systems.

Comfort the dying.

Relieve pain.

Tell the truth kindly.

If someone prays privately for strength, that is their choice. But as a society, we must not replace action with prayer or evidence with miracle stories.

Human beings have reduced suffering through medicine, public health, sanitation, vaccines, surgery, anesthesia, nutrition, emergency care, science, and social reform.

These are not small achievements.

They are the real miracles of human effort, though they are not supernatural.

They show what compassion can do when joined with knowledge.

## What Prayer and Miracles Reveal

Prayer and miracle claims reveal much about human beings.

They reveal fear.

They reveal hope.

They reveal our longing not to be alone.

They reveal our desire for justice, healing, protection, and meaning.

They reveal the emotional power of ritual and community.

They reveal how deeply we want love to be stronger than death.

These things deserve tenderness.

But they do not prove God.

When examined carefully, prayer does not show reliable divine response. Miracle stories appear across religions and often depend on testimony, coincidence, interpretation, and selective memory. Recovery follows medicine and biology. Disaster follows nature. Suffering continues despite sincere prayer.

The world looks as it would look if human beings pray, hope, interpret, and remember, but no personal God reliably intervenes.

That is the missing evidence.

## Chapter Summary

1. Prayer can be emotionally meaningful, but emotional comfort is not evidence of divine action.
2. If prayer worked reliably, clear patterns should appear beyond medicine, biology, chance, and social support.
3. The claim that God answers "yes, no, or wait" makes prayer impossible to test.
4. Miracle stories occur in many religions and cannot all prove conflicting beliefs.
5. Rare events, recoveries, and coincidences are not automatically supernatural.
6. Prayer can cause harm when failed prayer creates guilt or when prayer replaces action.

7. Human care, medicine, science, and practical help provide the clearest response to suffering.

## The Case So Far

The previous chapter examined divine hiddenness. This chapter asked whether prayer and miracles overcome that hiddenness. They do not. Prayer may comfort, and miracle stories may inspire, but neither provides reliable evidence that a personal God intervenes in the world.

## Key Takeaway

Prayer reveals human hope, fear, and longing, but it does not show reliable divine action. Miracles are usually ambiguous, selective, private, or explainable in natural ways. When suffering occurs, the strongest evidence of help comes not from heaven, but from human care, medicine, courage, and compassion.

## Sources and Further Reading


1. David Hume, **An Enquiry Concerning Human Understanding**.
2. Carl Sagan, **The Demon-Haunted World: Science as a Candle in the Dark**.
3. Michael Shermer, **The Believing Brain**.
4. Victor J. Stenger, **God: The Failed Hypothesis**.
5. James Randi, **The Faith Healers**.
6. Richard Dawkins, **The God Delusion**.
7. J. L. Mackie, **The Miracle of Theism**.


## Transition to the Next Chapter

If prayer and miracles do not provide clear evidence of divine action, another claim remains: perhaps God is known through the soul. Many religions teach that the real self is not the body, but an immortal spirit. The next chapter examines the mind, the brain, and whether the soul is still needed.

# Chapter 13: Mind Without Soul

## When the Person Changes

A family sits beside an elderly parent who no longer recognizes them.

The face is familiar. The hands are familiar. The voice may still sound the same. But something has changed. The mother who once remembered every birthday now forgets the names of her children. The father who once gave wise advice now repeats the same question again and again. A person who was gentle may become suspicious or angry. A person who was careful may become careless. A person who once told stories may fall into silence.

The family often says, with sadness, "He is still here, but he is not the same."

This experience is one of the most painful realities in medicine.

It also raises a deep question.

What is the self?

Religions often answer: the true self is the soul. The body may age, the brain may fail, but the soul remains the real person inside.

That idea is comforting.

But the evidence points in another direction.

When the brain changes, the mind changes. Memory, language, judgment, emotion, personality, awareness, and identity all depend deeply on the brain. The person we know does not float untouched above biology. The person is expressed through the living brain.

This does not make human beings less precious.

It makes them more fragile.

## The Ancient Idea of the Soul

The idea of the soul is very old.

Ancient people saw the difference between a living body and a dead body. One moment a person could speak, breathe, move, love, fear, and remember. After death, the body remained, but the person seemed gone. It was natural to imagine that something invisible had departed.

Dreams also encouraged belief in a soul. A sleeping person could appear to travel, meet the dead, or experience another world. Breath itself seemed mysterious. In many cultures, words for spirit, breath, wind, and life were closely connected.

Before neuroscience, the soul was an understandable explanation.

It seemed to explain life, consciousness, dreams, death, personality, and hope for survival after death.

But understandable does not mean true.

Many ancient explanations made sense before better evidence arrived. People once explained disease by spirits, storms by gods, and mental illness by possession. Those ideas were emotionally powerful, but natural explanations gradually replaced them.

The soul belongs to that older world of explanation.

Modern knowledge does not show a separate immortal self inside the body.

It shows a mind dependent on the brain.

## The Brain Is Not Just a Receiver

Some believers respond by saying the brain is only a receiver.

They compare it to a radio. If the radio is damaged, the sound becomes distorted, but the broadcast still exists somewhere else. In the same way, they say, brain damage may affect expression of the soul, but the soul itself remains untouched.

This analogy sounds comforting, but it does not fit the evidence well.

Brain damage does not merely distort a signal. It can remove specific abilities. A stroke in one area may impair speech. Damage in another area may affect vision. Disease in memory circuits may erase recent memories. Injury to frontal regions may change judgment, impulse control, and personality. Certain drugs can alter mood, perception, fear, pleasure, attention, and consciousness.

The changes are not random.

They follow brain structure and function.

If the brain were only a receiver for an independent soul, it is difficult to explain why specific damage produces specific mental losses. Why would injury to a small part of the brain remove the ability to name objects? Why would degeneration in certain networks erase memory? Why would chemicals change emotion and judgment? Why would anesthesia turn consciousness off?

The evidence points not to a receiver, but to an organ producing and organizing mental life.

The mind is what the brain does.

## Anesthesia and the Disappearance of Consciousness

Anesthesia is one of the clearest everyday challenges to belief in an independent soul.

A patient enters the operating room awake, aware, anxious, and speaking. Medication is given. Within moments, consciousness disappears. The patient does not simply relax. He does not merely become quiet. The stream of experience stops.

There is no memory of the surgery.

No awareness of time.

No sense of self observing from outside the body.

Then, as the drugs wear off, consciousness returns.

This is remarkable.

If the soul is the true conscious self, why can chemicals turn consciousness off so reliably? Why does awareness disappear when brain activity is altered? Why does the self return gradually as the brain returns to a waking state?

Anesthesia does not solve every philosophical mystery about consciousness. But it shows something powerful: conscious experience depends on the physical state of the brain.

The soul explanation adds little.

The medical explanation works.

Change the brain, and consciousness changes.

Alter the brain enough, and consciousness disappears.

## Stroke, Language, and the Fragile Self

A stroke can change a person in minutes.

A clot blocks blood flow. Brain tissue loses oxygen. Neurons are injured or die. Suddenly, a person may be unable to speak, understand speech, move one side of the body, recognize objects, read, write, or control emotions.

These losses are often specific.

A patient may know what he wants to say but cannot produce words. Another may speak fluently but without meaning. Another may understand language but be unable to name a common object. Another may ignore one side of space. Another may lose emotional control.

This is not what we would expect if language, memory, and identity came from an untouched soul.

It is exactly what we expect if mental abilities depend on brain networks.

The soul does not explain why damage to one region affects speech while damage to another affects movement or vision. Neuroscience does.

A stroke shows that the self is not a single invisible substance operating independently of the body. It is a complex living process supported by brain tissue, blood flow, oxygen, and neural connections.

The person is not less valuable because of this.

The person is more vulnerable than religion often admits.

## Dementia and the Vanishing of Memory

Dementia is one of the most difficult challenges to the soul idea.

Memory is central to personal identity. We know ourselves through our past: our childhood, family, language, work, habits, loves, losses, and stories. When memory fades, the person's world shrinks.

In Alzheimer's disease and other dementias, brain changes gradually damage memory, thinking, language, judgment, behavior, and personality. At first, the person may misplace objects or repeat questions. Later, he may forget familiar places, fail to recognize loved ones, lose words, become confused, or need help with basic activities.

The tragedy is not only that the body weakens.

The known person fades.

If an immortal soul carries the true self, why does damage to brain tissue erase the memories and traits that made the person recognizable? Why does the soul not preserve the mind clearly when the brain deteriorates? Why does the person's identity change as neurons and networks fail?

The brain-based explanation is painful, but clear.

Memory depends on the brain.

Identity depends on the brain.

Personality depends on the brain.

Dementia is not possession, divine punishment, or failure of soul. It is disease.

That understanding is medically and morally important.

It replaces blame with care.

## Personality Can Be Changed by the Brain

Many people imagine the soul as the seat of personality.

But personality can change when the brain changes.

Brain injury may make a calm person impulsive. A tumor may alter behavior. Damage to frontal brain regions can reduce judgment, restraint, empathy, or social awareness. Certain neurological conditions can produce suspicion, disinhibition, apathy, aggression, or emotional instability. Drugs, hormones, sleep deprivation, infection, pain, and metabolic problems can also change mood and behavior.

This is deeply important.

We like to think character is entirely spiritual or moral. But the brain plays a major role in how people feel, decide, control impulses, and relate to others.

This does not mean responsibility disappears. Human behavior is complex, and society still needs accountability. But it does mean that moral judgment should be informed by biology.

A person with brain disease is not simply "bad."

A person with depression is not spiritually weak.

A person with dementia is not choosing confusion.

A person with addiction, trauma, or severe mental illness may need treatment, structure, compassion, and accountability together.

The soul explanation often oversimplifies human behavior.

The brain explanation helps us understand and care more wisely.

## Near-Death Experiences

Near-death experiences are often presented as evidence for the soul.

People may report floating outside the body, seeing light, feeling peace, meeting loved ones, reviewing their life, or sensing another world. These experiences can be powerful and life-changing.

They should not be mocked.

But they do not prove that consciousness survives death.

Near-death experiences occur around extreme physical stress: lack of oxygen, altered blood flow, trauma, drugs, anesthesia, seizures, cardiac arrest, or recovery from critical illness. The brain under stress can produce vivid experiences. Memory can be shaped afterward. Cultural expectations can influence interpretation.

People from different cultures often interpret such experiences differently. A Christian may see Jesus or heaven. A Hindu may interpret the experience through Hindu imagery. A secular person may describe light, peace, or detachment without religious meaning.

The experience may be real as experience.

The interpretation may not be correct.

Most importantly, near-death is not the same as final, irreversible death. Reports come from people who survived and whose brains later resumed enough function to remember and describe something.

Near-death experiences show that the brain can produce powerful states near crisis.

They do not demonstrate an immortal soul.

## Split-Brain and the Divided Mind

Some neurological findings challenge the idea of one simple, indivisible soul.

In certain patients with severe epilepsy, the connection between the two brain hemispheres has been surgically cut to reduce seizures. These split-brain cases have revealed surprising features of consciousness.

Under special testing conditions, one hemisphere may know something the other cannot report. One hand may respond to information that the speaking hemisphere does not consciously recognize. The mind can appear divided in ways ordinary experience hides.

These cases are complex and should not be overstated.

But they raise a serious question.

If the soul is one indivisible center of consciousness, why can consciousness and awareness be divided by changing the brain's connections?

The brain-based view expects mental unity to depend on neural integration.

When integration is altered, mental unity can change.

The soul view has difficulty explaining why a supposedly simple spiritual self is so affected by the physical wiring between brain hemispheres.

Again, the evidence points toward mind as a brain process.

## The Soul Explains Less Than the Brain

The soul idea tries to explain many things: consciousness, identity, memory, morality, personality, and survival after death.

But when examined closely, it explains less than it promises.

It does not explain why memory disappears with dementia.

It does not explain why speech can vanish after stroke.

It does not explain why personality changes after brain injury.

It does not explain why anesthesia removes consciousness.

It does not explain why drugs alter thought and emotion.

It does not explain why specific brain regions support specific mental functions.

It does not explain why brain development shapes childhood, maturity, and aging.

The brain explains these facts far better.

This does not mean neuroscience has solved every mystery. Consciousness remains difficult. The relationship between brain activity and subjective experience is still studied and debated.

But an unsolved question is not evidence for a soul.

The soul is not an explanation. It is a label placed over mystery.

The brain is where the evidence points.

# Why We Feel Like More Than Matter

Many people resist a brain-based view because they feel like more than matter.

This feeling is understandable.

We do not experience ourselves as neurons firing. We experience thoughts, love, grief, memory, beauty, fear, decision, and meaning. A person is not felt from the inside as chemistry. Life is lived as experience.

But the fact that experience feels rich does not prove that it is supernatural.

Music is more than air pressure to the listener, but it still depends on physical vibration, ears, nerves, brain, memory, and emotion. A sunset is more than wavelengths to the person watching it, but it still depends on light, atmosphere, eyes, and brain. Love is more than chemicals in the way it is lived, but it still depends on the body and brain.

Natural does not mean worthless.

Physical does not mean meaningless.

To say that the mind depends on the brain is not to deny the depth of human life. It is to place that depth within nature.

The brain is matter organized in an extraordinary way.

That should increase wonder, not reduce it.

# Death Without a Soul

The soul is comforting because it promises survival.

If the body dies but the soul continues, death is not final. Loved ones may be reunited. Justice may be completed. The self may live on.

It is easy to understand why this belief is powerful.

But comfort is not evidence.

Everything we know about mind points to dependence on the brain. When the brain is injured, the mind is injured. When the brain degenerates, the mind degenerates. When the brain is deeply anesthetized, consciousness disappears. When the brain permanently stops functioning, there is no good evidence that the person continues as a conscious self.

This is painful.

But it can also make life more honest.

If the mind is mortal, then our time with one another is precious. If memory can fade, then love should be expressed now. If consciousness depends on the brain, then protecting health, reducing injury, caring for patients, and supporting the vulnerable become even more important.

A finite mind is not meaningless.

It is precious because it is temporary.

# The Moral Value of a Brain-Based Mind

A brain-based understanding of mind has moral power.

It reduces shame.

Mental illness is not demonic possession. Dementia is not moral failure. Epilepsy is not spiritual attack. Depression is not simply weakness. Addiction is not only bad character. Personality change after injury is not merely sin.

Human behavior still involves responsibility, but responsibility must be joined with understanding.

A natural view of mind supports medicine, therapy, rehabilitation, prevention, social support, and compassion. It helps families understand that a loved one with dementia is not choosing forgetfulness. It helps patients understand that depression is not proof of spiritual failure. It helps society respond to brain injury, trauma, and mental illness with more wisdom.

The soul idea may comfort some people.

But when used wrongly, it can blame people for conditions rooted in brain and body.

A scientific view does not remove dignity.

It protects it.

# Mind Without Soul, Humanity Without Loss

If there is no immortal soul, what remains?

Everything human remains.

Love remains.

Memory remains while the brain can hold it.

Music remains.

Grief remains.

Joy remains.

Kindness remains.

Moral responsibility remains.

Wonder remains.

The beauty of a human life does not depend on being supernatural. A person is not precious because an invisible essence has been inserted into the body. A person is precious because he or she is a conscious, feeling, thinking being capable of suffering, love, fear, hope, memory, relationship, and meaning.

The mind does not need a soul to matter.

The self does not need immortality to be real.

A natural life can still be profound.

In fact, understanding the mind as fragile may make us more tender toward one another. We are not indestructible spirits temporarily wearing bodies. We are living beings whose inner worlds depend on delicate organs, blood flow, oxygen, chemistry, development, and care.

That should humble us.

It should also deepen compassion.

## Chapter Summary

1. The ancient idea of the soul was understandable before neuroscience, but modern evidence points strongly to the brain.
2. Consciousness, memory, language, personality, judgment, and identity change when the brain changes.
3. Anesthesia, stroke, dementia, brain injury, drugs, and neurological disease all show the dependence of mind on brain.
4. Near-death experiences are powerful, but they do not prove survival after permanent brain death.
5. The soul explains less than neuroscience.
6. A brain-based view does not reduce human dignity. It makes human life more fragile and deserving of care.
7. The mind can be natural and still precious.

## The Case So Far

The previous chapter showed that prayer and miracles do not provide reliable evidence of divine action. This chapter examined another major religious claim: the soul. The evidence shows that the mind is deeply dependent on the brain, making an immortal independent soul unnecessary as an explanation.

## Key Takeaway

The mind does not need a soul to be real, deep, or valuable. Memory, personality, language, consciousness, and identity depend on the brain. When the brain changes, the person changes. This does not make human beings less precious. It makes them more fragile, more natural, and more deserving of compassion.

## Sources and Further Reading


1. Antonio Damasio, **Descartes' Error**.
2. Antonio Damasio, **Self Comes to Mind**.
3. Oliver Sacks, **The Man Who Mistook His Wife for a Hat**.
4. Eric R. Kandel, **In Search of Memory**.
5. Michael S. Gazzaniga, **The Consciousness Instinct**.
6. Stanislas Dehaene, **Consciousness and the Brain**.
7. Patricia S. Churchland, **Touching a Nerve: The Self as Brain**.
8. V. S. Ramachandran, **Phantoms in the Brain**.


## Transition to the Next Chapter

If the mind does not require a soul, another religious claim remains: morality requires God. Many people fear that without divine command, right and wrong collapse. The next chapter examines why morality can stand on human suffering, empathy, fairness, and reason without needing God.

# Chapter 14: Morality Without God

## The Child Who Helps

A small child sees another child fall on the playground.

No scripture is opened. No priest speaks. No doctrine is discussed. No fear of hell enters the moment. The child simply sees pain. Something in him responds. He walks over, reaches out, and tries to help.

That simple act tells us something important.

Morality does not begin with theology.

It begins with the recognition that another being can suffer.

Before commandments, there is pain. Before doctrine, there is fear. Before scripture, there is hunger, grief, injury, loneliness, and need. Human beings are moral not because a voice from heaven first told us to care, but because we live among other conscious beings whose lives can be helped or harmed.

Religion often claims ownership of morality. Many believers fear that without God, everything becomes permitted. They ask, "If there is no God, why be good?" or "Without divine command, how can anything be truly right or wrong?"

These are serious questions.

But they have answers.

Morality does not need God. It needs suffering, empathy, fairness, reason, responsibility, and the recognition that other lives matter.

## The Fear Behind the Question

The fear that morality will collapse without God is understandable.

Many people were taught from childhood that God is the source of right and wrong. They were told that commandments come from heaven, conscience is God's voice, and moral order depends on divine judgment. If that belief weakens, they may feel that the whole moral structure of life is in danger.

They may ask:

Who decides what is right?

Why should anyone be honest?

Why should the strong not crush the weak?

Why should cruelty be condemned?

These questions do not prove God.

They show that human beings need a moral foundation.

Religion offers one answer: obey God.

A secular moral view offers another: reduce suffering, protect dignity, promote well-being, respect fairness, and use reason to guide human life.

The second answer is not empty.

In many ways, it is more direct.

It does not begin by asking what God commands. It begins by asking who is harmed, who is helped, what is fair, and what kind of life allows conscious beings to flourish.

## Does Morality Need a Lawgiver?

Many believers argue that morality requires a divine lawgiver.

They say laws need lawmakers. Therefore, moral laws need God.

But this comparison is too simple.

Human laws are created because human beings live together and need order, safety, fairness, and protection from harm. We do not forbid murder merely because a ruler wrote a sentence in a book. We forbid murder because murder destroys a conscious life, spreads fear, breaks trust, and harms families and communities.

The wrongness of cruelty is not created by command.

It is recognized through the reality of suffering.

If a child is tortured, the act is wrong before anyone opens a scripture. If a person is enslaved, degraded, or starved, the harm is real before any religious authority comments on it. If a patient is denied care because of caste, religion, race, gender, or poverty, the injustice is visible in human suffering.

A lawgiver may command what is moral.

But the command is not what makes suffering matter.

Suffering already matters because conscious beings experience it.

# The Euthyphro Problem

This question is ancient.

Is something good because God commands it, or does God command it because it is good?

If something is good only because God commands it, then morality becomes arbitrary. If God commanded cruelty, would cruelty become good? If God commanded oppression, would oppression become moral? Most thoughtful believers would say no.

But if cruelty would still be wrong even if commanded, then morality does not depend entirely on God's command.

On the other hand, if God commands something because it is already good, then goodness exists as a standard that can be understood apart from God's command.

Either way, divine command does not solve the problem.

It only moves the question.

A moral rule is not good simply because someone claims it came from God. We must still ask whether it reduces suffering, respects persons, protects the vulnerable, and supports human flourishing.

That means morality must be able to judge religion.

It cannot be completely dependent on religion.

## Scripture Is Not a Perfect Moral Guide

Religious believers often point to scripture as the foundation of morality.

But scripture is morally mixed.

Sacred texts may contain compassion, wisdom, poetry, discipline, charity, humility, and concern for justice. They may also contain tribal violence, harsh punishments, unequal treatment of women, acceptance of slavery, caste or purity ideas, intolerance of outsiders, and commands that modern moral conscience rightly questions.

Most religious people do not follow every command in their scriptures literally.

They select, interpret, soften, ignore, explain, or reinterpret. They emphasize love and mercy while setting aside passages that appear cruel or outdated. This may be morally good, but it shows that believers are already using moral judgment outside the text to decide which parts of the text should guide them today.

That judgment comes from human conscience, empathy, history, reason, social progress, and moral reflection.

If scripture were the perfect source of morality, it would not need so much correction by later human moral development.

The fact that we judge scripture morally shows that morality is larger than scripture.

## Religion Has Inspired Good and Harm

A fair discussion must admit that religion has inspired much good.

Many believers have cared for the sick, fed the poor, comforted the grieving, opposed injustice, built schools, served refugees, and lived with courage and kindness. Their goodness should be respected. Their compassion is real.

But good actions by believers do not prove that God exists.

They prove that believers can be good.

Religion has also inspired or defended harm. It has been used to justify war, persecution, slavery, caste hierarchy, gender inequality, hatred of outsiders, punishment of doubt, fear of hell, and resistance to science. Not every religious person supports these harms, but religion has often provided sacred protection for them.

This mixed record matters.

If religion were the unique source of morality, its history should be morally clearer. Instead, religion reflects the human beings who create, interpret, and practice it. Sometimes it lifts people toward compassion. Sometimes it traps them inside fear, obedience, hierarchy, and tribal loyalty.

Religion can motivate morality.

It does not own morality.

## Morality Begins With Suffering

The simplest foundation of morality is this:

Conscious beings can suffer.

Pain is real. Fear is real. Hunger is real. Grief is real. Humiliation is real. Loneliness is real. Love is real. Trust is real. Injury is real.

A child being abused is not wrong merely because a religion says so. It is wrong because a conscious being is being harmed. A hungry person matters because hunger hurts. A patient gasping for breath matters because distress is real. A person treated as less than human matters because dignity and well-being are being damaged.

This is not mere opinion.

It is grounded in the reality of conscious experience.

We may disagree about some moral questions, but we do not begin from nothing. We begin from the fact that our actions affect beings who can feel. From there, reason asks: what reduces harm? What protects dignity? What allows trust? What promotes well-being? What rules can people live under fairly?

Morality begins where suffering begins.

No God is required for that.

## Empathy and the Moral Imagination

Empathy is the ability to feel, imagine, or understand another person's experience.

It is not perfect. Human beings can be selfish, biased, tribal, and cruel. We often feel more for those close to us than for strangers. We may ignore suffering when it is distant or hidden. That is why empathy alone is not enough.

But empathy is a beginning.

It breaks the prison of the self. It allows us to say, "That could be me. That could be my child. That person's pain matters too."

Moral progress often requires expanding empathy beyond narrow boundaries: beyond family, tribe, religion, caste, race, nation, gender, class, and species.

Religion sometimes helps this expansion, but it can also restrict it by dividing the world into believers and unbelievers, pure and impure, saved and damned, chosen and rejected.

Reason helps empathy grow wider.

It asks us to be consistent. If my suffering matters, why not yours? If my child deserves protection, why not another child? If my freedom matters, why not the freedom of someone born into another religion, family, caste, or nation?

This widening circle does not require God.

It requires imagination, fairness, and courage.

## Cooperation, Fairness, and Trust

Human morality also grows from social life.

Human beings are not solitary creatures. We are born helpless. We survive through care. We learn language from others. We depend on trust, cooperation, memory, teaching, and shared rules. A society without honesty, fairness, restraint, and concern cannot function well.

This helps explain why moral instincts appear in human life.

We value fairness because unfairness damages trust.

We value honesty because lies make cooperation dangerous.

We value kindness because life is vulnerable.

We value justice because unchecked harm destroys community.

These are natural facts about social beings.

They do not require divine command.

Other social animals show early forms of cooperation, care, attachment, and fairness. Human morality is far more complex because we have language, memory, culture, law, imagination, and reason. But the roots of moral life are not supernatural.

They grow from living together.

God is not needed to explain why social beings develop moral concern.

## Reward and Punishment Are Not the Highest Morality

Many religions teach moral behavior through reward and punishment.

Do good and receive heaven, merit, divine favor, blessing, good karma, or salvation. Do evil and receive hell, punishment, bad karma, divine anger, or spiritual loss.

Reward and punishment can influence behavior. Human societies also use them through law, parenting, education, and social approval.

But they are not the highest form of morality.

A person who avoids murder only because he fears hell has not reached deep morality. A person who helps the poor only to gain heavenly reward is doing good, but the motive is still incomplete. The highest moral response is not "What will happen to me?" but "What is happening to the other person?"

A child in pain should be helped because the child is in pain.

A hungry person should be fed because hunger hurts.

A victim should be protected because violence destroys.

A dying person should be comforted because fear and loneliness matter.

Mature morality moves beyond self-interest.

It acts because others matter.

That kind of morality does not need divine reward. It needs compassion and understanding.

# Atheists, Believers, and Human Character

Atheists can be kind, honest, loving, generous, disciplined, courageous, and deeply moral. They can be good parents, friends, physicians, nurses, teachers, scientists, judges, neighbors, artists, and citizens. They can care for the sick, protect children, oppose injustice, and live responsibly.

Believers know this in practice, even when theology tells them otherwise.

Many people have atheist friends or relatives whom they trust. They accept medical care from atheist doctors. They respect atheist scientists. They work with atheist colleagues. They rely on laws and institutions that do not require religious belief from every person involved.

This does not mean atheists are automatically good.

They are human. They can be selfish, cruel, dishonest, arrogant, or violent, just as believers can.

Belief in God also does not guarantee goodness.

A person may pray and still lie. A person may worship and still exploit others. A person may quote scripture and still hate. A person may believe in heaven and hell and still abuse power.

The truth is simple: morality depends on character, empathy, upbringing, reflection, accountability, social conditions, and personal responsibility.

Belief may influence morality.

It is not the source of all morality.

Disbelief does not prevent goodness.

## Moral Progress Often Required Questioning Religion

Many moral advances came not from simple obedience to tradition, but from questioning it.

Human societies have gradually challenged slavery, caste oppression, religious persecution, denial of women's rights, child marriage, cruelty to prisoners, torture, racism, and hatred of outsiders. Religious people sometimes helped these movements, and that should be acknowledged. But religious texts and institutions were also often used to resist them.

Moral progress usually required a deeper principle than "our tradition says so."

It required asking:

Who is being harmed?

Who is being treated as less than human?

Whose suffering has been ignored?

Whose voice has been silenced?

Are our inherited beliefs just?

This is moral reasoning, even when religious people use it.

It begins with human dignity and suffering, not blind obedience.

When religious morality improves, it often improves because human conscience pushes religion forward.

That matters.

Morality is not simply received from the past.

It is refined through human struggle.

## Is Secular Morality Just Opinion?

Believers sometimes say that without God, morality becomes only personal opinion.

But morality is not as empty as that.

Suffering and flourishing are real features of conscious life. Some actions reliably harm people. Some conditions reliably improve human well-being. Cruelty, neglect, humiliation, starvation, oppression, and violence damage lives. Care, honesty, safety, education, fairness, freedom, and love generally help lives.

These facts give morality a real foundation.

A person may say, "I do not care about suffering." But that does not make suffering unreal. It shows a failure of moral concern.

Medicine offers a useful comparison. Health is complex. Doctors may disagree about difficult cases. Cultures may value some outcomes differently. Yet health is not meaningless. Suffocation is worse than breathing. Starvation is worse than nourishment. Severe infection is worse than recovery.

Morality is similar.

It is complex, but not empty.

God is not needed to make suffering matter.

## The Physician's View of Morality

Medicine teaches morality every day.

A patient arrives in pain. The first question is not, "What religion is this patient?" The first question is, "What is wrong, and how can we help?"

The body does not become more worthy of care because it belongs to a believer. A child does not deserve less treatment because of the family's religion. An atheist patient does not deserve less compassion. A prisoner does not lose all human dignity. A confused elderly person is still a person.

The moral duty comes from vulnerability.

The patient can suffer. The patient can fear. The patient can be harmed or helped. That is enough.

In the hospital, morality is not an abstract doctrine. It is practical: relieve pain, tell the truth, respect dignity, protect privacy, obtain consent, prevent harm, care for the weak, and act with competence.

None of this requires belief in God.

It requires humanity.

## Morality Without Heaven

Without heaven, morality becomes more urgent, not less.

If there is no guaranteed afterlife where wrongs are corrected, then justice matters here. If there is no divine judge who will balance every account, then human justice matters more. If there is no eternal reward for the suffering, then reducing suffering now becomes more important.

Religion sometimes postpones justice to another world.

Atheism brings responsibility back to this one.

The hungry need food now.

The sick need care now.

The oppressed need justice now.

The lonely need companionship now.

The child needs protection now.

This does not make morality smaller.

It makes morality immediate.

Without heaven, human beings cannot leave compassion for later.

## A Better Question

The best moral question is not, "What does God command?"

The better question is:

Who is suffering?

Who is being harmed?

Who is being helped?

What is fair?

What protects dignity?

What reduces cruelty?

What allows conscious beings to flourish?

This way of thinking does not solve every moral problem instantly. Hard questions remain. Human beings will still disagree. Evidence must be studied. Consequences must be considered. Rights must be protected. Freedom and safety must be balanced. Justice and mercy must be held together.

But this approach keeps morality where it belongs: in the real lives of beings who can suffer and care.

Morality without God is not emptiness.

It is maturity.

It asks us to stop looking upward for commands and start looking honestly at one another.

## Chapter Summary

1. Morality does not begin with theology. It begins with the reality of suffering and care.
2. Divine command does not solve the problem of morality because we still need a standard for judging commands.
3. Scripture is morally mixed and often requires human moral judgment to interpret or correct it.
4. Religion has inspired good, but it has also defended harm.
5. Empathy, cooperation, fairness, reason, and human vulnerability provide natural foundations for morality.
6. Atheists can be moral, and belief in God does not guarantee goodness.
7. Without heaven, justice and compassion become more urgent in this life.

## The Case So Far

The previous chapter showed that the mind does not require a soul. This chapter showed that morality does not require God. Moral concern can grow from suffering, empathy, fairness, reason, responsibility, and the shared vulnerability of conscious beings.

## Key Takeaway

God is not needed for morality. Cruelty is wrong because suffering is real. Kindness is good because conscious beings can be helped. Fairness matters because people deserve dignity. Religion can inspire goodness, but it can also protect harm. A mature morality does not need divine command, fear of hell, or promise of heaven. It needs compassion guided by reason.

## Sources and Further Reading


1. Plato, **Euthyphro**.
2. David Hume, **An Enquiry Concerning the Principles of Morals**.
3. John Stuart Mill, **Utilitarianism**.
4. Peter Singer, **The Expanding Circle**.
5. Frans de Waal, **Primates and Philosophers**.
6. Steven Pinker, **The Better Angels of Our Nature**.
7. A. C. Grayling, **The God Argument**.


## Transition to the Next Chapter

If morality does not require God, another question remains: why do religions exist so widely and powerfully? If they are not divine revelations, what are they? The next chapter examines religion itself and asks why religions look so much like human creations.

# Chapter 15: Why Religions Look Human-Made

## The Accident of Birth

A child is born.

If he is born in a Hindu family, he is likely to learn Hindu prayers, festivals, rituals, stories, gods, and ideas about karma and rebirth.

If he is born in a Muslim family, he is likely to learn Islamic prayer, the Quran, Ramadan, mosque life, and belief in Allah as the one God.

If he is born in a Christian family, he is likely to learn about Jesus, the Bible, church, sin, salvation, heaven, and hell.

If he is born in a Buddhist family, he may learn about suffering, compassion, meditation, karma, rebirth, and liberation without belief in a creator God.

If he is born in a secular family, he may grow up without religious belief at all.

This simple fact is one of the strongest clues about religion.

Most people do not choose their first religion after carefully studying all religions and comparing the evidence. They inherit it. Religion arrives through family, language, food, festivals, songs, fear, love, authority, community, and childhood memory.

This does not mean every believer is thoughtless.

Many believers are sincere, intelligent, kind, and serious. Many think deeply about their faith. Many live morally beautiful lives.

But sincerity does not erase the pattern.

Religious belief follows birthplace, family, and culture far more than we would expect if one clear God had revealed one clear truth to all humanity.

It follows exactly the pattern we would expect if religions were human cultural systems.

## Religion Usually Comes Before Examination

For most people, religion is not first encountered as a claim to be tested.

It is encountered as belonging.

A child is told what "we" believe. He learns which book is sacred, which rituals matter, which prayers are important, which foods are allowed, which doubts are dangerous, and which outsiders are mistaken.

Before the child can compare evidence, religion has already become familiar.

Familiarity is powerful.

What is repeated in childhood begins to feel natural. What is shared by family begins to feel loving. What is taught by elders begins to feel authoritative. What is connected with weddings, funerals, festivals, grief, and identity becomes emotionally difficult to question.

This is why leaving religion can feel like betraying family, not merely changing an opinion.

Religion survives not only because people argue for it.

It survives because it becomes part of the self.

A belief learned early, repeated often, surrounded by love, protected by fear, and reinforced by community can feel like truth long before evidence is examined.

That is a human process.

## The Geography of God

If religion came mainly from clear divine revelation, geography should matter less.

But geography matters enormously.

Maps of religion often resemble maps of history, migration, conquest, empire, trade, family structure, language, and political power. Religions spread through parents, teachers, rulers, missionaries, merchants, marriage, colonization, social pressure, and sometimes violence.

This does not mean every religious spread was forced. Many people adopted religions voluntarily. Many found comfort, moral structure, beauty, or community in them.

But the broad pattern remains human.

Religions do not spread like one obvious truth recognized equally everywhere. They spread like cultures. They have centers, borders, histories, conflicts, reforms, sects, and local variations.

If one loving God had provided one clear universal revelation, we might expect humanity to converge more reliably on that truth.

Instead, humanity has produced many religions, each shaped by place and history.

The geography of belief looks less like divine clarity and more like human inheritance.

## Gods Often Resemble Their Societies

Religions often carry the values, fears, and social structures of the societies that produce them.

Warrior societies often imagine warrior gods.

Agricultural societies often develop rituals around rain, harvest, fertility, and seasonal cycles.

Hierarchical societies often imagine divine hierarchy.

Patriarchal societies often give men religious authority and place women under control.

Tribal societies often divide humanity into insiders and outsiders, pure and impure, chosen and rejected.

Ancient moral codes often reflect ancient social conditions.

This is what we would expect if human beings shape gods in their own image.

The gods may be described as eternal, but their concerns often look very local: food rules, purity rules, clothing rules, marriage rules, caste rules, gender rules, tribal loyalty, priestly authority, temple sacrifice, ritual obedience, and punishment of outsiders.

A universal God should rise far above local human prejudice.

Yet religions often preserve the prejudices of their time.

This does not mean every religious teaching is harmful. Many teachings are compassionate and wise. But the pattern strongly suggests human origin.

Religion often reflects humanity upward and then calls the reflection divine.

# Scriptures Have Human Fingerprints

Sacred texts are often treated as if they came from above history.

But when we read them carefully, they look deeply embedded in history.

They use human languages. They reflect ancient cultures. They contain local geography, political conflict, tribal memory, ritual law, poetry, myth, moral struggle, fear, hope, and authority. They address the problems of their time: kingship, war, purity, inheritance, marriage, food, temple practice, social order, and group survival.

They also contain moral limitations.

Some passages express compassion, justice, humility, love, and wisdom. Others accept or command things modern moral conscience rightly questions: slavery, harsh punishment, inequality, violence, patriarchy, caste-like hierarchy, or intolerance of outsiders.

Most believers do not live by every passage literally.

They interpret, soften, ignore, explain, or reject troubling parts. Often this is morally good. But it shows that human moral judgment is being used to filter scripture.

If scripture were perfect divine revelation, it should not need so much correction by later human conscience.

The more we study sacred texts historically, the more they look like human documents: powerful, influential, poetic, sometimes wise, sometimes troubling, and always shaped by the people who produced, preserved, translated, and interpreted them.

Human fingerprints are everywhere.

## Many Religions, Many Contradictions

Religions do not merely offer different styles of worship.

They make conflicting claims about reality.

One God. Many gods. No creator God.

One life. Rebirth. Heaven. Hell. Liberation. Annihilation.

Salvation by faith. Salvation by works. Liberation by knowledge. Freedom through meditation. Divine grace. Ritual purity. Devotion. Obedience.

One holy book. Another holy book. Many sacred texts. No final scripture.

One prophet. Many prophets. No prophet. Gurus, sages, saints, monks, priests, mystics, or enlightened teachers.

These differences are not small.

They concern the nature of reality, the meaning of life, the fate of the dead, the source of morality, and the path to salvation or liberation.

If one God clearly revealed truth, why does revelation look so confused?

Each religion can explain why it is right and others are wrong. But that is exactly what human groups do when defending inherited identity.

The existence of many sincere, conflicting religions is not strong evidence for God.

It is strong evidence that human beings create religious systems to answer human needs.

## Religion and Power

Religion is rarely only about private belief.

It often becomes connected to power.

Priests, clergy, gurus, monks, imams, rabbis, prophets, kings, caste leaders, elders, and institutions may claim authority over truth, morality, ritual, marriage, family, education, sexuality, death, and social order.

When authority is believed to come from God, questioning it becomes dangerous.

A political ruler can be opposed. A human law can be debated. A scientific claim can be tested. But a sacred command is harder to challenge because it is protected by divine authority.

This is one reason religion has often been useful to power.

It can teach obedience. It can sanctify hierarchy. It can define insiders and outsiders. It can make social arrangements appear eternal. It can turn human rules into divine rules.

This does not mean every religious leader is corrupt.

Many are sincere and compassionate.

But the structure is risky.

Whenever human beings claim to speak for God, human power becomes harder to question.

That is a very human pattern.

## Fear, Hope, and the Need for Meaning

Religion answers fear.

Fear of death. Fear of disease. Fear of disaster. Fear of punishment. Fear of meaninglessness. Fear of being alone. Fear that injustice will never be corrected. Fear that loved ones are gone forever.

These fears are real.

Religion offers powerful answers: heaven, rebirth, divine justice, forgiveness, protection, miracles, sacred meaning, and reunion after death.

Religion also gives hope.

It tells the poor that their suffering matters. It tells the dying that death is not the end. It tells the grieving that separation is temporary. It tells the lonely that someone unseen is present. It tells the wronged that justice will come.

These hopes can be emotionally powerful.

They explain why religion can feel deeply meaningful even when its supernatural claims are unsupported.

A fair atheist should acknowledge this.

Religion is not only ignorance, control, or fear. It is also a human attempt to survive pain, build community, express wonder, and carry hope across generations.

But emotional usefulness does not prove truth.

A story can help people endure without being literally true.

A ritual can strengthen community without opening heaven.

A belief can reduce fear and still be false.

Religion reveals human longing.

It does not necessarily reveal God.

## Ritual, Community, and Belonging

Religions are strong because they do more than make claims.

They create communities.

People gather, sing, pray, fast, feast, marry, mourn, celebrate, confess, give, serve, and remember together. Children grow up surrounded by repeated practices. Families mark time through religious calendars. Communities support one another during illness, birth, marriage, death, and crisis.

This social power helps religion survive.

A person who doubts a doctrine may still remain because religion is tied to family, friendship, identity, and belonging. Leaving may mean disappointing parents, troubling a spouse, confusing children, losing community, or becoming socially isolated.

That emotional cost is real.

It also shows that religion spreads through human bonds, not only through evidence.

People do not usually remain religious because they have carefully compared all arguments. They remain because religion is woven into life.

This makes religion powerful.

It does not make it true.

Human beings need belonging. Religion often supplies it. But belonging is not proof of divine revelation.

It is proof that human beings need one another.

## Religions Change Over Time

If religions came directly from a perfect God, we might expect them to be morally and intellectually complete from the beginning.

But religions change.

They reinterpret scripture. They adjust moral teaching. They respond to science. They soften old punishments. They revise views on slavery, caste, monarchy, gender, religious freedom, medicine, mental illness, and human rights. They divide into sects. They reform. They modernize. They resist change, then later accept what they once condemned.

This pattern looks human.

A perfect revelation should not need centuries of correction by human conscience, science, and social progress.

When science showed that Earth was not the center, religion adjusted.

When evolution explained life's diversity, many religious thinkers adjusted.

When modern medicine explained disease and mental illness, religion slowly adjusted.

When human rights movements challenged old hierarchies, many religious communities adjusted.

Religions often change because reality and moral progress pressure them to change.

That is not how divine perfection should behave.

It is how human institutions behave.

## Religion Resists Correction

Science changes when evidence requires change.

Not easily. Not perfectly. Scientists are human, and scientific communities can resist new ideas. But the method of science allows correction. A claim can be challenged by observation, experiment, replication, and better explanation.

Religion often works differently.

Religious claims may be protected by sacred authority. If evidence challenges doctrine, believers may reinterpret evidence, reinterpret scripture, or declare the matter beyond human reason. Doubt may be treated as sin. Critics may be accused of pride, rebellion, or spiritual blindness.

This resistance to correction is a major sign of human-made belief.

A claim that cannot be corrected by evidence becomes dangerous. It may survive not because it is true, but because questioning it has been forbidden.

This is why religious error can last for centuries.

If a belief is protected by fear, identity, and authority, it can outlive evidence.

Science is not perfect because scientists are not perfect.

But science has a built-in method for correction.

Religion often has a built-in method for preservation.

That difference matters.

## Why Religion Feels Sacred

If religions look human-made, why do they feel sacred?

Because human life contains sacred-feeling moments.

Birth, death, love, grief, guilt, forgiveness, awe, music, silence, sacrifice, and community can all feel larger than ordinary life. Religion gathers these experiences and gives them symbols, stories, rituals, places, and names.

A funeral can feel sacred because love and loss are deep.

A prayer can feel sacred because the heart is exposed.

A temple, church, mosque, shrine, or monastery can feel sacred because generations have poured emotion into that place.

A song can feel sacred because it carries memory, longing, and identity.

Sacred feeling is real as feeling.

But sacred feeling does not prove supernatural truth.

Human beings can create sacred meaning through memory, beauty, suffering, architecture, music, ritual, and shared emotion. A place can feel holy because people have made it holy through reverence, repetition, and love.

The sacred may reveal the depth of human experience.

It does not automatically reveal God.

## Religion as a Human Answer to Human Need

When seen clearly, religion looks like a human answer to human need.

Human beings fear death, so religions promise survival.

Human beings suffer injustice, so religions promise final justice.

Human beings feel guilt, so religions offer forgiveness.

Human beings fear chaos, so religions offer cosmic order.

Human beings need belonging, so religions create community.

Human beings face mystery, so religions offer explanation.

Human beings seek moral rules, so religions give commandments.

Human beings want hope, so religions tell stories of rescue.

This does not mean religion is foolish.

It means religion is deeply human.

It arose because human beings are vulnerable, imaginative, social, moral, fearful, hopeful, and aware of death. Religion is one of humanity's oldest attempts to make existence bearable.

But a human answer to a human need is not necessarily a divine revelation.

It may be emotionally powerful and still be invented.

## Why This Supports Atheism

Religion looks exactly like something human beings would create.

It follows geography.

It begins in childhood.

It reflects culture.

It carries ancient moral assumptions.

It supports identity.

It uses ritual and authority.

It answers fear and hope.

It changes over time.

It resists correction.

It produces many conflicting systems.

This pattern is difficult to reconcile with one clear revelation from a loving God.

It fits much better with naturalism.

Human beings created gods because they needed explanations, comfort, control, community, moral order, and hope. Then they inherited those gods, defended them, revised them, fought over them, and called them eternal.

The more we understand religion historically and psychologically, the less divine it looks.

It looks human.

## Respecting Believers Without Accepting Belief

To say religion looks human-made is not to insult believers.

Human beings create many meaningful things: music, poetry, medicine, law, art, philosophy, family traditions, memorials, ceremonies, and moral ideals. Something can be human-made and still be emotionally important.

But truth claims must still be examined.

A religious ritual may comfort people. A scripture may contain wisdom. A community may support the sick. A believer may live with kindness and courage. These things deserve respect.

But they do not prove God.

The question is not whether religion matters to people. It clearly does.

The question is whether religion reveals a supernatural reality.

The evidence points elsewhere.

Religion reveals humanity: our fear, imagination, hope, morality, need for belonging, struggle with death, and desire for meaning.

That is enough to explain why religion exists.

## Chapter Summary

1. Most people inherit religion through birthplace, family, language, culture, and childhood.
2. Religious belief follows geography and history more than clear universal revelation.
3. Gods, scriptures, rituals, and moral codes often reflect the societies that produced them.
4. Religions provide comfort, hope, community, identity, and answers to fear.
5. Religions change over time and often require correction by science, morality, and human progress.

6. The pattern of religion fits human creation better than divine revelation.
7. Respecting believers does not require accepting unsupported religious claims.

## The Case So Far

The previous chapter showed that morality does not require God. This chapter examined religion itself. Instead of looking like one clear message from a divine source, religions look like human cultural systems shaped by geography, childhood, fear, hope, authority, power, and history.

## Key Takeaway

Religions look human-made because they behave like human-made systems. They are inherited through childhood, shaped by culture, divided by geography, tied to power, filled with human hopes and fears, and changed by history. Religion may comfort, inspire, and organize people, but it does not prove God. It explains humanity more than it reveals divinity.

## Sources and Further Reading


1. Daniel C. Dennett, **Breaking the Spell: Religion as a Natural Phenomenon**.
2. Pascal Boyer, **Religion Explained**.
3. Émile Durkheim, **The Elementary Forms of Religious Life**.
4. David Sloan Wilson, **Darwin's Cathedral**.
5. James George Frazer, **The Golden Bough**.
6. Michael Shermer, **The Believing Brain**.

7. Richard Dawkins, **The God Delusion**.

## Transition to the Next Chapter

If the universe does not need God, life does not need a designer, mind does not need a soul, morality does not need divine command, and religion looks human-made, one question remains: what happens when we let God go? The next chapter turns from argument to life. It asks what we gain when we stop defending unsupported belief and begin living fully in the real world.

# Chapter 16: What We Gain When We Let God Go

## The Fear of Losing God

Many people do not hold on to God only because of arguments.

They hold on because they are afraid of what might happen if belief disappears.

Will life lose meaning?

Will morality collapse?

Will death become unbearable?

Will love become temporary and therefore less real?

Will suffering become empty?

Will the universe feel cold?

Will family and community be lost?

Will guilt and fear remain even after belief is gone?

These fears are real.

For many people, religion is not just an idea. It is childhood, family, festivals, prayers, funerals, weddings, food, memory, identity, and belonging. Losing belief may feel not like changing an opinion, but like losing a home.

That emotional difficulty should be respected.

But fear does not decide truth. And losing belief in God does not mean losing everything good.

Many things people associate with religion do not actually belong to religion alone: love, compassion, gratitude, beauty, community, forgiveness, moral courage, wonder, service, and hope.

When God is let go, these do not disappear.

They can become more honest.

## The First Freedom: Thinking Without Fear

One of the first things many people gain is freedom from religious fear.

Fear of hell.

Fear of divine punishment.

Fear of bad karma.

Fear of questioning.

Fear of offending God.

Fear of thinking the wrong thought.

Fear of dying with the wrong belief.

Fear that doubt itself is a sin.

These fears can enter the mind early and remain for years. A person may no longer believe intellectually, yet still feel fear emotionally. Childhood teaching can leave deep tracks.

Letting go of God does not always remove fear in one day. But it begins the process.

A frightening claim still needs evidence. Hell is not true because it is terrifying. Divine punishment is not real because it was taught early. Fear may shake the body, but it does not prove the belief.

When religious fear weakens, the mind can breathe.

A person can ask questions without trembling. He can examine scripture without guilt. He can admit doubt without feeling wicked. He can live ethically without imagining invisible surveillance.

This is not rebellion.

It is mental freedom.

# Freedom From False Guilt

Religion often creates guilt beyond ordinary moral responsibility.

Natural desires may be called sinful. Doubt may be called pride. Illness may be treated as punishment. Poverty may be explained as karma. Grief may be judged as lack of faith. A person may grow up feeling watched, measured, and morally stained before life has even truly begun.

Some guilt is useful.

If we harm someone, guilt can help us repair the damage. It can guide us back toward honesty, kindness, and responsibility.

But false guilt is different.

False guilt makes people ashamed for being human.

Letting go of God can remove guilt that never belonged there. A person can stop feeling sinful for thinking. A patient can stop wondering what spiritual failure caused illness. A grieving parent can stop asking whether God is punishing the family. A doubter can stop feeling morally dirty for needing evidence.

Science-based atheism does not remove responsibility.

It removes imaginary guilt.

Real morality becomes clearer when false guilt fades.

# Freedom From Blaming the Innocent

One of the moral gains of atheism is freedom from cruel explanations.

If there is no divine plan behind disease, then the sick are not being punished.

If there is no karma governing poverty, then poverty is not proof of moral failure from a past life.

If there is no God sending disasters, then victims are not targets of divine anger.

If there is no demon causing mental illness, then patients need care, not spiritual blame.

If there is no sacred reason for a child's cancer, then the child's suffering does not need to be defended as part of a higher plan.

This matters deeply.

Religious explanations often try to protect God by placing meaning on suffering. But sometimes the most compassionate thing we can say is simple:

This should not have happened.

The person did not deserve it.

Now we must help.

Atheism allows suffering to be seen without theological distortion. It does not ask us to call tragedy a blessing, punishment, test, or mystery. It allows us to respond with medicine, justice, comfort, and action.

That is not cold.

It is humane.

## Freedom From Sacred Authority

When people claim to speak for God, ordinary human authority can become dangerous.

A leader may say, "This is not my opinion. This is God's command." A scripture may be used to silence questions. A tradition may be treated as beyond criticism. A child may be taught obedience before understanding. A woman may be told her place is divinely assigned. A caste, class, tribe, or religious group may be declared superior by sacred order.

Letting go of God weakens this kind of power.

No human being gets automatic authority by claiming divine support. No book becomes morally perfect because it is called holy. No tradition becomes immune to examination because it is old. No leader should be obeyed without question because he claims to represent heaven.

This does not mean all religious authority is abusive.

Many religious leaders are kind and sincere.

But sacred authority is risky because it makes human power harder to challenge.

A free mind asks for reasons.

A humane society asks for evidence.

A moral community allows questions.

Without God, authority must answer to human well-being.

## Meaning Without Heaven

Many people fear that without God, life becomes meaningless.

But meaning does not come only from eternity.

A song is meaningful though it ends. A meal with family is meaningful though it lasts an hour. A child's laughter is meaningful though it fades into memory. A flower is beautiful though it wilts. A life can be meaningful though it is finite.

Meaning arises from relationship, love, work, service, creativity, learning, courage, beauty, and the reduction of suffering.

A physician comforting a patient is meaningful.

A parent caring for a child is meaningful.

A teacher opening a mind is meaningful.

A scientist discovering truth is meaningful.

A friend sitting beside the grieving is meaningful.

None of this requires heaven.

In fact, finiteness may deepen meaning. If this life is the only life we know, then this day matters. This conversation matters. This kindness matters. This chance to help matters.

Eternity is not required for value.

Sometimes the shortness of life is exactly what makes it precious.

## Love Without Eternity

Religion often promises that love will continue forever.

That promise is emotionally powerful because love does not want to end. We want the dead restored. We want reunion. We want parents, spouses, children, and friends to remain. We want love to defeat death.

Atheism cannot honestly promise that.

But it does not make love less real.

Love matters while it exists. A mother's care is not meaningless because she is mortal. A friendship is not false because it ends. A marriage is not worthless because death separates the partners. The love given to a child remains part of that child's life. The kindness given to a patient remains real even if both people eventually die.

Temporary does not mean meaningless.

A beautiful sunset is temporary. Music is temporary. Childhood is temporary. Health is temporary. Yet all can matter deeply.

Atheism asks us not to reduce love to a promise about another world.

It asks us to love more carefully in this one.

## Morality Without Reward

When God is let go, morality does not disappear.

It matures.

We no longer need to ask, "What reward will I receive?" or "What punishment will I avoid?" We can ask a better question:

"What does this action do to another being who can suffer?"

This is the heart of secular morality.

Feed the hungry because hunger hurts.

Tell the truth because trust matters.

Protect children because they are vulnerable.

Oppose cruelty because pain is real.

Support justice because power without fairness crushes the weak.

Care for the sick because illness frightens and weakens human beings.

Morality becomes less about spiritual accounting and more about real consequences.

Without heaven, goodness is not a transaction.

Without hell, compassion is not forced by fear.

We do good because others matter.

That is enough.

## Wonder Without Worship

Some people fear that atheism removes wonder from the world.

It does not.

The natural world is astonishing without being supernatural. The atoms in our bodies were formed in ancient stars. Life on Earth shares a deep ancestry. The brain can think about the universe from which it emerged. A newborn develops from a single cell. A tree turns sunlight into living tissue. A bird crosses continents. The human mind can create music, mathematics, medicine, poetry, and love.

None of this becomes smaller without God.

It may become larger.

Instead of seeing the world as a stage built for humans by divine command, we see ourselves as part of a vast natural story. We are not the center of the universe. We are a brief, conscious part of it.

That humility can deepen wonder.

A rainbow is not less beautiful because we understand light. A flower is not less beautiful because we understand evolution. The stars are not less beautiful because we understand physics.

Understanding does not destroy wonder.

It purifies it.

## Death Without Illusion

The hardest loss for many people is the loss of belief in life after death.

Atheism does not soften this with false promises. Death is real. Conscious life appears to depend on the brain. When the brain permanently stops, the person as we know him appears to end.

This truth can be painful.

But it can also make life more honest.

If there is no guaranteed second life, then we should not waste this one. If there is no heavenly repair for cruelty, then justice matters here. If loved ones are mortal, then we should love them now. If our time is limited, then resentment, hatred, and needless cruelty become even more tragic.

Death gives life urgency.

It teaches us to say what should be said, repair what can be repaired, forgive when possible, protect health, reduce suffering, and not postpone love indefinitely.

Atheism does not make death easy.

It makes life more precious because death is real.

## Community Without Dogma

Leaving God does not mean leaving community forever.

It may mean losing one community, and that can be painful. Some families and religious groups make doubt difficult. Some people may feel isolated when belief changes. This emotional cost should not be minimized.

But community can be rebuilt.

Human beings can gather around compassion, learning, art, science, service, friendship, shared meals, music, nature, books, volunteering, public health, justice, and care for children and the elderly.

A community does not need dogma to be meaningful.

It needs trust.

It needs kindness.

It needs shared responsibility.

Religious community often works not because its supernatural claims are true, but because people need one another. That need remains after God is gone.

Atheists and secular people should take this seriously. Humans do not live by arguments alone. We need belonging, ritual, memory, grief support, celebration, and shared purpose.

A godless life should not be a lonely life.

It should be a more honest human life.

## Responsibility Without Rescue

One of the greatest changes after letting go of God is the return of responsibility.

If no God is guaranteed to rescue the suffering, human beings must act.

If no divine judge is guaranteed to correct injustice, human beings must build justice.

If no prayer is guaranteed to heal disease, human beings must support medicine and research.

If no sacred plan explains disaster, human beings must prepare, protect, rebuild, and learn.

If no heaven is guaranteed, human beings must make this life more livable.

This can feel heavy.

But it is also empowering.

We are not helpless children waiting for a hidden parent. We are conscious beings capable of care, knowledge, cooperation, courage, and moral progress. We cannot fix everything, but we can fix some things. We cannot prevent all suffering, but we can reduce much of it.

The absence of God does not make action meaningless.

It makes action necessary.

## Peace Without Pretending

There is a quiet peace that can come from no longer defending what one does not believe.

A person no longer has to explain why God allowed a child to die. He no longer has to say a disaster was part of a plan. He no longer has to pretend that unanswered prayer was secretly answered. He no longer has to force ancient scriptures to fit modern conscience. He no longer has to call doubt a sin.

He can say:

I do not know everything.

I will follow evidence.

I will not blame the innocent.

I will not call suffering good.

I will love without needing eternity.

I will be moral without needing heaven.

I will find meaning in this life.

This is not despair.

It is honesty.

And honesty can bring peace.

## What We Keep

Letting go of God does not require throwing away every human value religion carried.

We can keep compassion.

We can keep gratitude.

We can keep community.

We can keep music.

We can keep silence.

We can keep moral seriousness.

We can keep care for the poor.

We can keep reverence for life.

We can keep ceremonies that honor love, grief, birth, death, and memory, if we reshape them honestly.

We can keep the human needs that religion tried to answer, while letting go of unsupported supernatural claims.

We do not need to keep fear of hell, shame of doubt, blame of the sick, sacred inequality, divine surveillance, or obedience to unquestioned authority.

We can separate the human from the supernatural.

We can keep what heals and leave what harms.

## A Fuller Human Life

A life without God can be full.

It can contain love, family, friendship, work, service, curiosity, laughter, grief, beauty, moral courage, intellectual honesty, and deep wonder. It can be disciplined, generous, joyful, and compassionate. It can be serious about truth and tender toward suffering.

It can also be imperfect, as every human life is.

Atheism does not make people automatically wise. It does not remove selfishness, fear, anger, tribalism, or error. It is not a magic cure for the human condition.

But it clears away one major illusion: that the universe is governed by a personal God who will explain, judge, rescue, punish, reward, and repair all things.

Once that illusion is gone, we can begin with reality.

Reality is enough to demand our attention.

Suffering is enough to demand our compassion.

Life is enough to demand our gratitude.

Human beings are enough to demand our moral concern.

No God required.

## Chapter Summary

1. Many people fear losing God because religion is tied to family, meaning, morality, death, and identity.
2. Letting go of God can bring freedom from fear, false guilt, victim-blaming, divine surveillance, and sacred authority.
3. Atheism does not remove meaning, love, morality, wonder, or community.
4. A finite life can still be deeply meaningful.
5. Without belief in divine rescue, human responsibility becomes more urgent.
6. A godless life can be honest, compassionate, moral, beautiful, and full.

## The Case So Far

The previous chapters showed why God is not needed to explain the universe, life, body, suffering, mind, morality, or religion. This chapter turned from argument to life. It showed that letting go of God is not only a loss of belief. It can also be a gain in freedom, honesty, responsibility, and compassion.

## Key Takeaway

Letting go of God does not mean losing morality, meaning, love, wonder, or hope. It means giving up fear, false guilt, divine surveillance, victim-blaming, and unsupported claims. A life without God can be honest, humane, responsible, and deeply meaningful. We do not need heaven to love, scripture to be moral, or divine rescue to care for one another.

## Sources and Further Reading


1. A. C. Grayling, **The God Argument**.
2. Bertrand Russell, **The Conquest of Happiness**.
3. Carl Sagan, **Pale Blue Dot**.
4. Paul Kurtz, **The Courage to Become**.
5. Richard Dawkins, **Unweaving the Rainbow**.
6. Sam Harris, **Waking Up**.
7. Julian Baggini, **Atheism: A Very Short Introduction**.


## Transition to the Next Chapter

The journey of the book is now almost complete. We have examined faith, arguments for God, the universe, life, the body, suffering, prayer, mind, morality, religion, and the life that remains when God is no longer needed. The final chapter will bring the entire case together and answer the book's central question one last time: is God required?

# Conclusion: No God Required

## Returning to the Final Question

This book began with a simple question.

Is God required?

Not as a word for nature. Not as a poetic name for love, mystery, or beauty. But as the personal God of religion: the God who creates, designs, reveals, commands, judges, protects, heals, answers prayer, gives souls, and gives meaning to suffering.

If such a God exists, reality should show it clearly.

The world should not look exactly as it would look if no such God existed.

This book has followed that question across the major areas where God is usually placed: faith, arguments, the universe, life, the body, suffering, prayer, mind, morality, religion, meaning, and death.

The pattern has been steady.

Faith feels powerful, but it is not evidence.

Arguments for God raise serious questions, but they do not prove the God of religion.

Science has repeatedly replaced supernatural explanations.

The universe inspires awe, but it does not reveal a loving personal ruler.

Life appears designed, but evolution explains that appearance.

The body is brilliant, but fragile and imperfect.

Disease follows biology, not sin.

Suffering does not behave like justice.

Prayer does not show reliable divine intervention.

The mind depends deeply on the brain.

Morality grows from suffering, empathy, fairness, and reason.

Religion looks shaped by human history, culture, fear, hope, power, and belonging.

Meaning, love, wonder, and compassion remain fully possible without God.

Each point matters by itself.

Together, they form the cumulative case.

The world behaves like nature.

No God is required.

## The Wound Religion Cannot Heal

The strongest challenge to the personal God is not merely scientific.

It is moral.

Innocent suffering remains the wound that religion cannot fully heal.

A child with cancer.
A newborn with a genetic disease.
A family destroyed by earthquake.
An animal suffering before human beings existed.
A patient losing memory to dementia.
A parent praying beside a bed where medicine, not miracle, is doing the visible work.

Religion tries to explain such suffering in many ways: sin, karma, test, mystery, divine plan, free will, soul-making, or hidden purpose.

But these answers do not satisfy.

Disease does not follow guilt. Natural disaster does not follow moral worth. Viruses do not respect innocence. Genetic mutations do not wait for sin. Animals suffered long before human beings could make moral choices. Prayer does not reliably rescue the suffering.

A world governed by perfect love should not look like this.

A natural world would.

This is not a cold conclusion. It is a humane one.

It tells the sick they are not cursed. It tells the poor they are not paying for past-life sins. It tells the grieving they do not have to call tragedy a blessing. It tells the suffering child that no theology is needed to justify pain.

Suffering should not be defended.

It should be reduced.

That is where science-based atheism becomes morally clearer than religious explanation.

It does not protect God from suffering.

It protects the sufferer from blame.

## The Silence That Remains

If God wanted to be known, the world should be clearer.

But sincere seekers often find silence.

Religions divide by geography. Scriptures conflict. Doctrines contradict. Miracles remain uncertain. Prayer produces no reliable pattern. Believers interpret events through the faith they inherited. Honest doubters are often blamed for not seeing what is not clear.

A loving God should not be so easy to confuse with absence.

If belief matters deeply, evidence should not be hidden behind culture, childhood, fear, and interpretation. If God wants relationship, he should not be less clear than a human parent, teacher, or friend.

The silence of God is not a small problem.

It fits the natural explanation.

There is no clear divine voice because no personal God is speaking.

## What Science Has Shown

Science has not solved every mystery.

It has not fully explained the origin of the universe, the origin of life, consciousness, or every detail of reality. Honest science admits this.

But incomplete knowledge is not evidence for God.

Mystery is not a doorway through which religion can safely enter and claim victory. History teaches caution. Again and again, mysteries once filled by gods became fields of natural knowledge.

Lightning became electricity.

Disease became biology.

Mental illness became brain, body, trauma, and environment.

Life's diversity became evolution.

The body became anatomy, physiology, genetics, and medicine.

The mind became increasingly linked to the brain.

This does not make reality smaller.

It makes reality more understandable.

Science does not remove wonder. It removes false certainty. It teaches us to say, "We do not know yet," instead of pretending that an ancient answer has solved the question.

That humility is not weakness.

It is honesty.

## Mind, Morality, and Meaning

Religion often says that without God, the deepest parts of life collapse.

The soul disappears. Morality disappears. Meaning disappears. Hope disappears.

But that fear is mistaken.

The mind does not need a soul to be real. Memory, personality, love, grief, imagination, and consciousness are natural, but not worthless. Their dependence on the brain makes them fragile, not meaningless.

Morality does not need divine command. Cruelty is wrong because suffering is real. Kindness is good because conscious beings can be helped. Fairness matters because power can crush the weak. Honesty matters because trust is necessary. Compassion matters because life is vulnerable.

Meaning does not need heaven. A finite life can still contain love, service, learning, beauty, courage, friendship, and care. A song matters though it ends. A flower matters though it fades. A human life matters because it is felt from within and connected to others.

Temporary does not mean false.

Natural does not mean empty.

Human does not mean imaginary.

God is not needed for the things that make life precious.

## What We Lose and What We Keep

Letting go of God can involve real loss.

Some people lose certainty. Some lose childhood comfort. Some lose a sense of being watched over. Some lose belief in reunion after death. Some lose community. Some feel grief, fear, or loneliness.

That should not be mocked.

Belief can be deeply woven into memory, family, culture, and identity.

But something is gained too.

We gain freedom from fear of hell.

Freedom from false guilt.

Freedom from blaming the sick.

Freedom from calling disaster divine punishment.

Freedom from defending cruel scriptures.

Freedom from treating doubt as sin.

Freedom from sacred authority that cannot be questioned.

Freedom from pretending that silence is an answer.

And we keep what is most human.

We keep love.

We keep compassion.

We keep gratitude.

We keep wonder.

We keep moral responsibility.

We keep the duty to reduce suffering.

We keep music, memory, friendship, family, service, courage, beauty, and truth.

We keep the ability to stand beside the sick, feed the hungry, protect the weak, comfort the grieving, teach children, seek knowledge, and build a more decent world.

These things do not require God.

They require humanity.

## The Moral Beauty of Reality

There is moral beauty in letting go of unsupported belief.

A child's pain no longer needs to be justified as God's plan.

A woman's oppression no longer needs to be sanctified by scripture.

A poor person's hunger no longer needs to be explained by karma.

A patient's illness no longer needs to be called punishment.

A doubter's honesty no longer needs to be treated as rebellion.

A person's worth no longer depends on correct belief.

When God is removed from these explanations, compassion becomes clearer.

We stop asking how to defend God.

We start asking how to help people.

That shift matters.

It brings attention back to the only world where we know suffering occurs and the only world where help can be given.

This life.

This body.

This person.

This moment.

## The Final Answer

Is God required to explain the universe?

No.

Is God required to explain life?

No.

Is God required to explain the human body?

No.

Is God required to explain disease?

No.

Is God required to explain suffering?

No.

Is God required to explain the mind?

No.

Is God required to explain morality?

No.

Is God required to explain religion?

No.

Is God required for meaning, love, wonder, courage, or compassion?

No.

Again and again, the answer is the same.

No God required.

This does not make life small.

It makes life honest.

We are natural beings in a natural universe. We are fragile, temporary, conscious, feeling, thinking, loving animals. We suffer. We hope. We make mistakes. We seek meaning. We fear death. We build stories. We ask questions. We care for one another.

We are not fallen souls waiting for rescue.

We are human beings responsible for one another in the only world we know.

That is enough.

## Final Words

The purpose of this book has not been to make the reader angry at religion.

It has been to ask whether religion is true.

The purpose has not been to mock believers.

It has been to examine belief.

The purpose has not been to remove beauty, meaning, or morality.

It has been to show that none of these require God.

Reality is natural, but not empty.

Life is finite, but not meaningless.

Morality is human, but not imaginary.

Love is temporary, but not false.

Suffering is undeserved, but not sacred.

Human beings are fragile, but not worthless.

We can understand reality without the divine.

We can live morally without divine command.

We can face suffering without blaming the innocent.

We can create meaning without heaven.

We can love without eternity.

We can be fully human without God.

No God required.

## Conclusion Summary

1. The personal God of religion requires strong evidence, but faith, tradition, scripture, authority, and comfort do not provide enough.
2. Science explains the universe, life, body, disease, mind, and religion better than supernatural belief.
3. Innocent suffering, natural evil, divine hiddenness, and failed prayer strongly challenge belief in a loving, all-powerful God.
4. Morality, meaning, love, wonder, and compassion do not require God.
5. Science-based atheism is not emptiness. It is an honest, humane, evidence-based way to understand reality.

## Key Takeaway

The case for atheism is cumulative. The universe behaves naturally. Life evolved. The body is imperfect. Disease follows biology. Suffering is not just. Prayer lacks reliable evidence. The mind depends on the brain. Morality grows from suffering and compassion. Religion looks human-made. God is not needed to explain reality or to live a meaningful moral life. The honest conclusion is simple: no God required.

## Sources and Further Reading


1. Carl Sagan, **The Demon-Haunted World: Science as a Candle in the Dark**.
2. Bertrand Russell, **Why I Am Not a Christian**.
3. David Hume, **Dialogues Concerning Natural Religion**.
4. Charles Darwin, **On the Origin of Species**.
5. Richard Dawkins, **The Blind Watchmaker**.
6. Daniel C. Dennett, **Breaking the Spell: Religion as a Natural Phenomenon**.
7. J. L. Mackie, **The Miracle of Theism**.
8. J. L. Schellenberg, **Divine Hiddenness and Human Reason**.
9. Sean Carroll, **The Big Picture**.
10. A. C. Grayling, **The God Argument**.

# Appendix A: The Cumulative Case for Science-Based Atheism

## Why the Cumulative Case Matters

The strongest case for atheism does not rest on one argument alone.

It rests on a pattern.

One unanswered prayer may trouble a believer. One scientific discovery may weaken a religious claim. One example of innocent suffering may raise a moral question. One contradiction between religions may create doubt.

But the full force of science-based atheism appears when all these points are seen together.

The universe does not require God.

Life does not require a designer.

The body does not look perfectly designed.

Disease does not follow sin.

Suffering does not behave like justice.

Prayer does not show reliable divine intervention.

The mind does not require an immortal soul.

Morality does not require divine command.

Religion does not look like one clear revelation from heaven.

Each point matters. Together, they form a powerful conclusion: the world behaves exactly as we would expect if there were no personal God governing it.

This appendix gathers that cumulative case in one place.

## The God Being Examined

This book has not argued against every vague use of the word God.

If someone uses God to mean nature, love, mystery, energy, or the universe, then the discussion changes. Nature exists. Love exists. Mystery exists. The universe exists. Renaming these things God does not prove the personal God of religion.

The case in this book is directed at a stronger claim.

It examines the God said to create the universe, design life, give souls, reveal scripture, command morality, answer prayer, perform miracles, judge human beings, punish evil, reward faith, protect the innocent, and give meaning to suffering.

That God makes claims about reality.

Those claims should leave evidence.

If such a God existed and acted in the world, we would expect clearer signs than we find. We would expect prayer to show stronger patterns. We would expect divine justice to be visible. We would expect revelation to be less confused. We would expect innocent suffering to be better explained. We would expect mind to show independence from brain. We would expect religion to look less like geography, culture, and childhood inheritance.

Instead, the evidence points elsewhere.

## The Standard of Evidence

A claim as large as God requires strong evidence.

Faith is not enough. Tradition is not enough. Authority is not enough. Comfort is not enough. Personal feeling is not enough. A sacred text claiming divine authority is not enough. Mystery is not enough.

These may explain why people believe.

They do not prove that the belief is true.

In ordinary life, we already use this standard. We ask for evidence before accepting a medical treatment, a legal accusation, a scientific discovery, or a claim of danger.

The God claim should not be protected from the same care.

The burden of proof belongs to the person making the claim.

Atheism does not need to disprove every imagined supernatural being. It only needs to say that belief is not justified without sufficient evidence.

Science-based atheism adds something more: natural explanations fit reality better than religious ones.

## The Evidence Map

The cumulative case can be summarized in fifteen points.

1. **Faith is not reliable evidence.**
   Faith can support many conflicting religions. A method that can defend opposite beliefs cannot reliably prove truth.
2. **The strongest arguments for God fail.**
   First cause, fine-tuning, design, morality, consciousness, scripture, beauty, and religious experience raise serious questions, but they do not prove the personal God of religion.
3. **Science has repeatedly replaced supernatural explanations.**
   Lightning, eclipses, disease, seizures, mental illness, and life's diversity were once explained supernaturally. Natural explanations proved stronger.
4. **The universe does not need divine creation.**
   The cosmos is vast, ancient, lawful, mostly hostile to life, and not centered on human beings.

5. **Life does not need a designer.**
   Evolution explains adaptation, common ancestry, biological imperfection, extinction, and the appearance of design.
6. **The body is not perfectly designed.**
   The human body is brilliant but fragile, full of trade-offs, aging, disease, and evolutionary compromises.
7. **The origin of life remains a scientific question, not proof of God.**
   An unsolved problem is not supernatural evidence.
8. **Innocent suffering challenges a loving, all-powerful God.**
   Disease, disaster, genetic illness, and animal suffering do not behave like divine justice.
9. **Free will cannot explain natural evil.**
   Cancer, viruses, earthquakes, birth defects, genetic disease, and prehuman animal suffering are not caused by human choice.
10. **God remains hidden.**
    Sincere seekers often find silence, and religious belief follows geography, childhood, culture, and inherited identity.
11. **Prayer and miracles lack reliable evidence.**
    Prayer may comfort, but it does not show clear divine intervention. Miracle claims are usually ambiguous, selective, private, or explainable in natural ways.
12. **Mind depends on brain.**
    Stroke, dementia, anesthesia, drugs, injury, and disease change memory, personality, language, judgment, and consciousness.
13. **Morality does not require God.**
    Moral concern grows from suffering, empathy, fairness, cooperation, reason, and human vulnerability.

14. **Religion looks human-made.**
    Religions reflect culture, power, fear, hope, geography, ritual, identity, and history.
15. **Meaning does not require heaven.**
    Love, compassion, wonder, truth, courage, and responsibility remain fully human without God.

These points are not disconnected.

They are one pattern seen from many directions.

## What the Pattern Predicts

If the personal God of religion existed, we might expect the world to look different.

A loving God should make innocent suffering less blind.

A just God should make moral justice clearer in nature.

A God who answers prayer should leave stronger evidence of answered prayer.

A God who wants relationship should not be hidden from sincere seekers.

A God who reveals truth should not leave humanity divided by geography into conflicting religions.

A God who designed the body should not produce so many flaws, diseases, and painful compromises.

A God who gave an immortal soul should not make memory, personality, and consciousness so dependent on brain tissue.

A God who commands morality should provide moral guidance clearer than what human conscience later has to correct.

But the world does not match these expectations.

It matches naturalism better.

Nature is beautiful but indifferent.
Life is adaptive but wasteful.
The body is functional but flawed.
Disease is biological, not moral.
Prayer is comforting, not reliably effective.
Religion is meaningful, but human.
Morality is real because suffering is real.

The religious view must add many defenses: mystery, hidden purpose, tests, free will, karma, sin, soul-making, symbolic scripture, selective miracles, and unknowable plans.

Atheism does not need these defenses.

It simply says that reality behaves naturally because reality is natural.

## Why Science Is Stronger Than Supernatural Explanation

Science is not perfect.

It changes. It makes mistakes. It can be misused. Scientists are human. Knowledge remains incomplete.

But science has one great strength: it can correct itself.

A scientific claim can be tested, challenged, improved, or rejected. Germ theory replaced older ideas of disease because evidence supported it. Evolution replaced special creation because it explained life better. Neuroscience replaced many spiritual explanations of mind because brain evidence became overwhelming.

Science does not win because scientists are always right.

It wins because its method allows error to be exposed.

Religion often works differently. Sacred claims may be protected from correction. Doubt may be called sin. Tradition may be valued above evidence. Authority may preserve claims long after they fail. Mystery may be used to prevent questioning.

This difference matters.

A method that can correct itself is more trustworthy than a method that mainly protects itself.

## Why Suffering Is Central

The problem of suffering is not merely one argument among many.

It is the moral center of the case.

If God were only an abstract first cause, suffering might not matter in the same way. But religion usually claims far more. It claims that God is loving, wise, just, powerful, and involved in human life.

That claim collides with reality.

Children suffer without guilt. Animals suffered long before humans. Natural disasters strike blindly. Genetic diseases appear before moral choice. Cancer follows cells, not sin. Poverty follows social and economic conditions, not spiritual worth. Prayer does not reliably rescue the innocent.

A world governed by perfect love should not look like this.

A natural world would.

The atheist view does not make suffering painless. It does something morally important: it stops defending suffering as divine wisdom.

It allows us to say:

This is not deserved.
This is not sacred.
This should be reduced.

That is a clearer and more compassionate response than trying to protect God from the evidence of pain.

## Why Hiddenness Matters

A loving personal God should not be difficult to distinguish from no God.

If God wants human beings to know him, love him, obey him, and perhaps face eternal consequences for belief or disbelief, then clarity should matter.

Yet belief is deeply shaped by birthplace and childhood. Scriptures conflict. Religions disagree. Prayer is ambiguous. Miracles are uncertain. Sincere seekers often remain unconvinced. Honest nonbelief exists.

This is difficult for theism.

It is expected under atheism.

If religions are human creations, we would expect many religions. We would expect them to follow geography. We would expect them to reflect culture. We would expect people to experience the divine in forms they were taught. We would expect confusion, disagreement, reform, and conflict.

That is exactly what we see.

Divine hiddenness is not just silence.

It is the failure of the world to show the clear presence of the God religion describes.

## Why Mind Does Not Need Soul

The soul once seemed necessary to explain life, identity, consciousness, and death.

Today, the evidence points strongly to the brain.

Memory changes with brain disease. Personality changes with brain injury. Language can be lost after stroke. Consciousness can disappear under anesthesia. Drugs can alter mood and judgment. Dementia can slowly erase the person family members once knew.

These facts do not make human beings less valuable.

They make human beings more fragile.

The soul explains little. It does not explain why specific brain injuries produce specific mental losses. It does not explain why chemicals can turn consciousness off. It does not explain why memory fades when brain tissue deteriorates.

A brain-based mind explains these facts better.

Consciousness remains a deep question, but mystery does not justify belief in an immortal soul. The mind can be natural and still precious.

## Why Morality Does Not Need God

Morality begins with suffering.

A child in pain matters before any scripture is opened. A hungry person matters before any theology begins. A patient gasping for breath matters because distress is real. A person treated with cruelty is harmed whether or not a divine command has been spoken.

Morality grows from empathy, reason, fairness, cooperation, dignity, and the shared vulnerability of conscious beings.

God is not needed to make suffering matter.

Divine command also creates a problem. If something is good only because God commands it, morality becomes arbitrary. If God commands it because it is good, then goodness is not dependent on God's command.

Religious people can be moral. Atheists can be moral. Belief in God does not guarantee goodness, and disbelief does not prevent it.

A mature morality does not ask only, "What will God reward or punish?"

It asks, "Who is harmed, who is helped, and what kind of life allows conscious beings to flourish?"

## Why Religion Looks Human

Religion is one of the clearest human phenomena in history.

It answers fear, death, suffering, guilt, injustice, identity, mystery, and the need for belonging. It creates rituals, communities, symbols, stories, sacred places, moral rules, and shared memory.

That is why it is powerful.

But its power does not prove its truth.

Religions follow human patterns. They are inherited in childhood. They cluster by geography. They reflect culture and history. They protect authority. They change over time. They disagree deeply with one another. Their scriptures bear human fingerprints. Their moral teachings often improve only after pressure from human progress.

This is what we would expect if religions were made by human beings.

It is not what we would expect from one clear revelation by a perfectly wise God.

Religion explains human longing.

It does not prove divine reality.

## The Simplest Explanation

The best explanation is not always the one we prefer emotionally.

It is the one that fits the evidence with the fewest unnecessary additions.

Theism must explain why God is hidden, why prayer is unreliable, why suffering is blind, why religions conflict, why bodies are flawed, why disease is biological, why minds depend on brains, why moral progress often corrects scripture, and why God's actions are so difficult to distinguish from nature and chance.

Atheism has a simpler explanation.

There is no personal God governing reality.

The universe is natural. Life evolved. Bodies are imperfect because evolution works through compromise. Disease follows biology. Suffering has no built-in moral meaning. Prayer comforts but does not control nature. Mind depends on brain. Morality is a human response to suffering. Religion is a human creation.

This explanation is not always emotionally easy.

But it is clear.

It fits the evidence.

## What Science-Based Atheism Offers

Science-based atheism is not only a rejection.

It offers a way to live honestly.

It allows us to stop blaming the sick for sin, karma, or divine punishment.

It allows us to stop treating doubt as rebellion.

It allows us to stop defending suffering as part of a hidden plan.

It allows us to stop placing moral authority in ancient texts beyond correction.

It allows us to build morality from compassion and reason.

It allows us to value this life without needing another.

It allows us to feel wonder without worship.

It allows us to say "I do not know" without filling the silence with unsupported certainty.

It places responsibility where it belongs: in human hands.

Without God, we must care for one another more seriously, not less.

# Appendix Summary

1. The strongest case for atheism is cumulative, not dependent on one argument.
2. The personal God of religion makes claims that should leave clear evidence.
3. Faith, tradition, comfort, scripture, authority, and mystery are not enough.
4. Science explains reality better than supernatural belief.
5. Suffering, hiddenness, failed prayer, brain-based mind, and human-made religion strongly weaken the God claim.
6. Morality, meaning, love, and wonder remain possible without God.
7. The simplest explanation is that reality is natural and no personal God is governing it.

# Key Takeaway

The whole pattern of evidence points in one direction. The universe behaves naturally. Life evolved. The body is imperfect. Disease follows biology. Suffering is not just. God remains hidden. Prayer lacks reliable evidence. The mind depends on the brain. Morality begins with suffering and compassion. Religion looks human-made. The cumulative conclusion is clear: reality does not need God.

## Sources and Further Reading

1. Carl Sagan, **The Demon-Haunted World: Science as a Candle in the Dark**.
2. Bertrand Russell, **Why I Am Not a Christian**.
3. David Hume, **Dialogues Concerning Natural Religion**.
4. Charles Darwin, **On the Origin of Species**.
5. Richard Dawkins, **The Blind Watchmaker**.
6. Daniel C. Dennett, **Breaking the Spell: Religion as a Natural Phenomenon**.
7. J. L. Mackie, **The Miracle of Theism**.
8. J. L. Schellenberg, **Divine Hiddenness and Human Reason**.
9. Sean Carroll, **The Big Picture**.
10. A. C. Grayling, **The God Argument**.

## Transition to Appendix B

The cumulative case has now been gathered in one place. But readers may still hear common objections from family, friends, religious teachers, or their own lingering fears. The next appendix offers clear, calm replies to those objections.

# Appendix B: Common Religious Objections and Clear Replies

## How to Use This Appendix

When people begin questioning God, they often hear the same objections again and again.

"You cannot disprove God."
"Science cannot explain everything."
"Without God, there is no morality."
"What about miracles?"
"What if you are wrong?"

These questions can feel powerful, especially when they come from family, friends, religious teachers, or from one's own childhood fears.

This appendix offers clear, calm replies.

The purpose is not to win arguments or humiliate believers. The purpose is to think carefully. A good reply should be honest, respectful, and firm. It should separate emotional comfort from evidence. It should avoid personal attack. It should return again and again to the central question:

Is the claim true?

A belief may be comforting, ancient, popular, or deeply personal.

But truth still requires evidence.

## Objection 1: "You Cannot Disprove God"

This is true in one limited sense.

No one can disprove every possible idea of God, especially if God is defined vaguely as mystery, energy, love, or something beyond all testing. But that is not the same as proving that God exists.

Many things cannot be absolutely disproved. We cannot disprove every invisible spirit, every hidden demon, every undetectable fairy, or every imagined supernatural being. Yet we do not believe them without evidence.

The burden of proof belongs to the person making the claim.

Atheism does not need to prove that no possible God can exist in any possible sense. It only says that the evidence does not justify belief in the personal God of religion.

A clear reply is:

"I do not claim to disprove every imagined God. I am saying that the God described by religion has not been supported by strong evidence."

## Objection 2: "Science Cannot Explain Everything"

Science cannot explain everything.

That is true.

Science does not yet fully explain the origin of the universe, the origin of life, consciousness, or every detail of nature. Honest science admits uncertainty.

But "science does not know everything" does not mean "therefore God."

A gap in knowledge is not evidence for God. It is simply a gap in knowledge.

History teaches caution. Lightning, disease, eclipses, seizures, mental illness, and the diversity of life were all once explained through gods, spirits, curses, or supernatural forces. Natural explanations later replaced those ideas.

The honest answer to mystery is not unsupported certainty.

It is inquiry.

A clear reply is:

"Science does not know everything, but not knowing something yet does not prove God. A mystery is not evidence for a miracle."

## Objection 3: “Something Cannot Come From Nothing”

This objection is often used to argue that God must have created the universe.

But the question immediately returns:

Where did God come from?

If everything must have a cause, then God needs a cause. If God does not need a cause, then the rule has been changed to protect God. One cannot say everything needs a cause and then exempt the very being one is trying to prove.

Also, modern cosmology is not as simple as “something came from nothing.” The origin of the universe is a deep scientific question involving physics, space, time, energy, and laws we may not fully understand.

Even if the universe had a cause, that would not prove the personal God of religion. It would not prove scripture, prayer, heaven, hell, miracles, divine commands, or a loving plan.

A clear reply is:

“If everything needs a cause, then God needs a cause too. And even if the universe had a cause, that would not prove the God of religion.”

## Objection 4: “The Universe Is Fine-Tuned for Life”

The fine-tuning argument points to real questions.

The universe has physical constants that allow stars, chemistry, planets, and life. It is reasonable to wonder why the universe has these properties.

But fine-tuning does not prove God.

We do not know whether the constants could have been different, whether deeper laws explain them, whether multiple universes exist, or whether observers can only appear in universes compatible with observation.

There is also a deeper problem: the universe is not very friendly to life.

Most of it is empty space, radiation, extreme cold, burning stars, black holes, vacuum, and lifeless distance. Even on Earth, life survives through disease, predation, extinction, disaster, and struggle.

A universe designed by perfect love should look more life-centered than this.

A clear reply is:

"Fine-tuning raises a question, but it does not prove God. The universe is mostly hostile to life, and we do not yet know whether deeper natural explanations exist."

## Objection 5: "Life Is Too Complex to Evolve"

Life is complex.

But complexity alone does not prove design.

Evolution explains how simple beginnings can lead to complex forms over long periods through variation, inheritance, natural selection, mutation, genetic drift, and environmental pressure. Natural selection is not pure chance. Random variation is filtered by non-random survival and reproduction.

The appearance of design is exactly what evolution explains.

Eyes, wings, lungs, immune systems, and brains did not appear fully formed in one step. They developed through gradual changes, reuse of older structures, and long natural history.

The body also contains many flaws that fit evolution better than perfect design: back pain, choking risk, difficult childbirth, blind spots, wisdom teeth problems, appendix disease, cancer, aging, and genetic disorders.

A clear reply is:

"Life looks designed because natural selection can produce adaptation. Complexity does not prove a designer, especially when the design contains so many flaws."

## Objection 6: "Evolution Is Only a Theory"

In science, a theory is not a guess.

A scientific theory is a well-supported explanation that connects many lines of evidence. Germ theory, atomic theory, cell theory, and the theory of gravity are not casual guesses. They are powerful explanatory frameworks.

Evolution is supported by fossils, anatomy, genetics, embryology, biogeography, observed change, and DNA evidence. It explains why living things share ancestry, why bodies have imperfections, why species change, and why life forms fit into a branching tree.

Calling evolution "only a theory" misunderstands the scientific meaning of the word.

A clear reply is:

"In science, a theory is a strong explanation supported by evidence. Evolution is one of the best-supported ideas in biology."

## Objection 7: "DNA Is a Code, So There Must Be a Coder"

DNA is often described as a code because it stores biological information.

But "code" here is an analogy.

DNA is a molecule. It changes through mutation, recombination, duplication, selection, and inheritance. Its patterns can be studied naturally. Shared DNA, broken genes, genetic errors, and common ancestry all fit evolution.

Human codes require coders because they are products of human minds. DNA is not a written language in the human sense. It is chemistry shaped by reproduction and selection.

The code analogy can be useful, but it should not be turned into proof of God.

A clear reply is:

"DNA is called a code by analogy, but it is a molecule shaped by natural processes. Calling it a code does not prove a divine coder."

## Objection 8: "Without God, There Is No Morality"

Morality does not need God.

Cruelty is wrong because suffering is real. Kindness is good because conscious beings can be helped. Fairness matters because people can be harmed by injustice. Honesty matters because trust is necessary for human life.

If morality depends only on God's command, then we must ask: is something good because God commands it, or does God command it because it is good? If it is good only because God commands it, morality becomes arbitrary. If God commands it because it is already good, then goodness is not dependent on God's command.

Human moral life grows from empathy, suffering, fairness, reason, cooperation, and responsibility.

A clear reply is:

"God is not needed to make suffering matter. Cruelty is wrong because it harms conscious beings, not because a command had to make pain important."

## Objection 9: "If There Is No God, Why Be Good?"

This question assumes that goodness needs reward or punishment.

But mature morality is not only about what happens to us. It is about what happens to others.

We should help the hungry because hunger hurts. We should protect children because they are vulnerable. We should tell the truth because trust matters. We should oppose cruelty because suffering is real.

A person who avoids evil only because of fear of hell has not reached the highest moral level. A better reason to be good is that other beings matter.

Atheists can be moral, loving, responsible, and compassionate. Belief in God does not guarantee goodness, and disbelief does not prevent goodness.

A clear reply is:

"I try to be good because other people can suffer, not because I expect heavenly reward or fear divine punishment."

## Objection 10: "Without God, Life Has No Meaning"

Meaning does not require eternity.

A song is meaningful though it ends. A meal with family is meaningful though it is brief. A child's laughter is meaningful though the moment passes. A human life can be meaningful even if it is finite.

Meaning comes from love, relationship, service, creativity, learning, beauty, courage, and the reduction of suffering.

If this life is the only life we know, it becomes more precious, not less.

A clear reply is:

"Life does not need to last forever to matter. Love, kindness, work, beauty, and compassion can be meaningful within a finite life."

## Objection 11: "What About Miracles?"

Miracle stories exist in many religions.

Christians, Hindus, Muslims, Sikhs, Buddhists, Catholics, Protestants, gurus, saints, healers, and spiritual movements all report miracles. Since these religions often contradict one another, miracle stories cannot all prove their religions true.

Many miracle claims involve incomplete information, uncertain diagnosis, natural recovery, coincidence, selective memory, emotional interpretation, or testimony that cannot be checked.

This does not mean everyone reporting a miracle is lying.

It means extraordinary claims require strong evidence.

A clear reply is:

“Miracle stories occur in many religions and are often unclear or poorly tested. A powerful story is not the same as reliable evidence.”

## Objection 12: “Prayer Works”

Prayer may comfort people.

It may calm the mind, create hope, bring family together, and help people endure suffering. These effects can be real and meaningful.

But comfort is not the same as evidence that God changes external events.

If prayer worked reliably, we should see clear patterns beyond medicine, biology, chance, and social support. Instead, believers recover and die much like everyone else. Some prayers are followed by recovery. Many are not. If every outcome is called an answer, prayer cannot be tested.

A clear reply is:

"Prayer may help people emotionally, but it has not shown reliable evidence of changing reality through divine intervention."

## Objection 13: "I Personally Experienced God"

Personal experiences can be powerful.

A person may feel peace, presence, guidance, forgiveness, or love. Such experiences should not be mocked.

But personal experience needs interpretation. People in different religions have different experiences and interpret them through their own traditions. A Christian may feel Jesus. A Hindu may feel Krishna or Shiva. A Muslim may feel Allah. A Buddhist may feel awakening or compassion. A secular person may feel awe, peace, or unity without belief in God.

The experience may be real as experience.

The explanation may still be mistaken.

A clear reply is:

"I respect your experience, but people in many religions have powerful experiences. The feeling itself does not prove one specific God."

## Objection 14: "Scripture Proves God"

A scripture cannot prove itself by claiming to be divine.

Many religions have sacred texts. The Bible, Quran, Vedas, Upanishads, Guru Granth Sahib, Buddhist texts, and many others are treated as sacred by different communities. They cannot all be the final and complete revelation if they contradict one another.

Scriptures also show human features: language, history, culture, moral limitations, editing, translation, interpretation, and disagreement among believers.

A sacred text may contain wisdom, poetry, and moral insight.

But that does not prove divine authorship.

A clear reply is:

"A book claiming divine authority does not prove that it came from God. Many religions have sacred books, and they often disagree."

## Objection 15: "Many Intelligent People Believe in God"

That is true.

Many intelligent, educated, kind, and thoughtful people believe in God.

But intelligence does not guarantee that every belief is true. Smart people can inherit beliefs, defend identity, follow culture, fear doubt, value comfort, or reason differently about religion than they do about other claims.

Many intelligent people also reject belief in God.

The question is not whether some smart people believe.

The question is whether the belief is supported by evidence.

A clear reply is:

"Intelligent people can believe different things. The issue is not intelligence, but evidence."

## Objection 16: "Many Scientists Believed in God"

Some scientists have believed in God.

Some scientists today still do.

But this does not prove God. Scientists are human beings with culture, childhood, emotion, identity, and personal beliefs. A scientist's religious belief is not the same as scientific evidence for that belief.

Also, many past scientists lived in societies where religious belief was socially expected or legally safer. Some used the word God in a philosophical or poetic way, not always in the full religious sense of a prayer-answering personal deity.

Science should be judged by evidence, not by the private religious views of individual scientists.

A clear reply is:

"Some scientists believed in God, but that is not evidence for God. The question is what the evidence shows, not what any individual scientist believed."

## Objection 17: "Atheism Is Just Another Faith"

Atheism is not faith in the same sense as religion.

Atheism does not require belief in scripture, revelation, miracle, heaven, hell, soul, divine command, or unseen supernatural beings. It is the position that belief in God is not justified by the evidence.

Science-based atheism also accepts correction. If strong evidence for God appeared, the conclusion could be reconsidered.

Religious faith often asks people to believe without sufficient evidence or to preserve belief despite contrary evidence.

These are not the same.

A clear reply is:

"Atheism is not faith in another God. It is withholding belief because the evidence is not strong enough."

## Objection 18: "You Just Want to Sin"

This objection attacks motive instead of answering evidence.

Many people leave religion reluctantly. They may lose comfort, certainty, family peace, community, and belief in an afterlife. Many do not leave because they want an easier life. They leave because they cannot honestly believe anymore.

Also, atheists are not free from moral responsibility. They still live in families and societies. They still care about trust, love, law, fairness, reputation, and the suffering of others.

A clear reply is:

"My disbelief is not about wanting to sin. It is about not seeing enough evidence for the claim."

## Objection 19: "You Are Angry at God"

A person cannot be angry at a God he does not believe exists in the same way he can be angry at a real person.

Some atheists may be angry at religious harm, fear, hypocrisy, abuse, or false teaching. That is different from being angry at God.

Many atheists are not angry. They are simply unconvinced.

Calling atheism anger avoids the evidence question.

A clear reply is:

"I am not rejecting God because I am angry. I am rejecting the claim because I do not find the evidence convincing."

## Objection 20: "Religion Gives People Comfort, So Why Question It?"

Religion can comfort people.

That should be acknowledged.

It may help people grieve, face death, build community, practice gratitude, or endure hardship.

But comfort does not prove truth.

A false belief can comfort. A story can help people emotionally without being literally true. A child may feel comfort believing a monster cannot enter because of a magic blanket, but comfort does not make the blanket magical.

Also, some religious comfort comes with harm: fear of hell, guilt, victim-blaming, intolerance, control, shame, and resistance to medical or social progress.

A clear reply is:

"Comfort matters, but truth matters too. A belief can be comforting and still be false."

## Objection 21: "If Belief Helps People, What Is the Harm?"

Sometimes belief helps.

But beliefs also shape behavior.

If people believe illness is punishment, they may blame the sick. If they believe poverty is karma, they may ignore injustice. If they believe prayer is enough, they may delay treatment. If they believe one scripture is perfect, they may defend harmful commands. If they believe outsiders are damned or impure, they may justify discrimination.

Not every believer does these things.

But unsupported sacred beliefs can protect real harm.

Truth matters because beliefs have consequences.

A clear reply is:

"Belief can comfort, but it can also harm. We should examine beliefs because they affect how people treat one another."

## Objection 22: "You Need God to Explain Beauty and Wonder"

Beauty and wonder are real.

But they do not prove God.

A sunset, a mountain, music, mathematics, a newborn child, a bird in flight, or the night sky can create deep awe. Science does not destroy that awe. It can deepen it by showing the natural processes behind what we see.

A rainbow is not less beautiful because we understand light. A flower is not less beautiful because we understand evolution. The stars are not less beautiful because we understand physics.

Wonder does not require worship.

A clear reply is:

"I feel wonder too. But beauty is not proof of God. Nature can be deeply beautiful without being supernatural."

## Objection 23: "What If You Are Wrong?"

This is often called Pascal's Wager.

It suggests that believing in God is safer because if God exists, belief may be rewarded, and disbelief may be punished.

But the wager has serious problems.

Which God should one believe in? There are many religions with conflicting claims. Believing in the wrong God may not help. Also, sincere belief cannot simply be chosen like insurance. A person cannot honestly force himself to believe something he finds unsupported.

A just God should value honesty more than fearful pretending.

A clear reply is:

"I cannot make myself believe from fear. If God is just, honest doubt should be better than dishonest belief."

## Objection 24: "God Works in Mysterious Ways"

Some things are mysterious.

Human knowledge is limited.

But "mystery" should not be used to protect every claim from examination. If every problem is answered by mystery, then no evidence matters. Suffering becomes mystery. Failed prayer becomes mystery. Conflicting religions become mystery. Hiddenness becomes mystery. Cruel scriptures become mystery.

Mystery can explain anything, which means it explains very little.

A clear reply is:

"Mystery may be real, but it cannot be used to defend every unsupported claim. If a belief explains all outcomes, it explains nothing clearly."

## Objection 25: “You Still Cannot Explain Why There Is Something Rather Than Nothing”

This is a deep question.

Atheism does not pretend to have every answer.

But God does not solve the question as easily as people think. If God exists, then there is still something rather than nothing: God. We can then ask why God exists. Saying “God is necessary” is a claim, not a demonstrated answer.

It is more honest to admit mystery than to place God inside it without evidence.

A clear reply is:

“I do not know why there is something rather than nothing. But saying God exists does not remove the mystery. It only moves it to God.”

## Objection 26: “Life After Death Has Been Proven”

Life after death has not been proven.

Near-death experiences, visions, dreams of the dead, medium claims, and feelings of presence can be emotionally powerful. But they do not establish that consciousness survives permanent brain death.

The evidence from neuroscience points strongly in the other direction. Memory, personality, consciousness, and identity depend on the brain. When the brain is damaged, the mind changes. When the brain stops functioning permanently, there is no reliable evidence that the person continues as a conscious self.

A clear reply is:

"Near-death experiences and visions may feel powerful, but the evidence shows that mind depends on brain. Survival after death has not been proven."

## Objection 27: "Without Religion, Society Will Collapse"

Society depends on trust, law, cooperation, education, fairness, family, compassion, institutions, and shared responsibility.

Religion can support some of these, but it is not the only foundation for them.

Many secular societies function with strong public trust, human rights, science, medicine, education, and social order. Many religious societies have also had corruption, violence, inequality, and injustice. Religion does not automatically make a society moral, and secularism does not automatically destroy morality.

What matters is how people treat one another.

A clear reply is:

"Society does not depend only on religion. It depends on trust, justice, education, compassion, and responsible institutions."

## Objection 28: "Atheism Has Caused Great Evil Too"

Atheists, like believers, can do terrible things.

No worldview should deny human cruelty.

Some political regimes that rejected religion committed great harm. But their crimes do not prove God exists, just as religious wars, persecution, caste oppression, witch hunts, terrorism, or abuse do not prove atheism true.

The question is whether the evil was caused by atheism itself or by authoritarian power, ideology, nationalism, fear, propaganda, and human cruelty.

Atheism simply means lack of belief in God. By itself, it does not command violence, worship a leader, create a political system, or justify oppression.

A clear reply is:

"Atheists can do wrong, and believers can do wrong. Human cruelty is real. But that does not prove God, and it does not make unsupported religious claims true."

## Objection 29: "You Are Taking Away Hope"

A difficult truth is not the same as hopelessness.

Atheism may remove certain hopes: heaven, divine rescue, cosmic justice, reunion after death. That loss can be painful.

But it leaves other hopes.

Hope for healing through medicine.
Hope for justice through human courage.
Hope for comfort through love.
Hope for progress through knowledge.
Hope for meaning through service.
Hope for less suffering through compassion and action.

These hopes are not imaginary. They are grounded in human effort.

A clear reply is:

"I am not against hope. I am against false certainty. We can build real hope through love, knowledge, justice, and compassion."

## Objection 30: "Why Not Just Believe Anyway?"

Because truth matters.

Belief is not only a private feeling. It shapes how we see disease, suffering, morality, death, children, outsiders, science, and human responsibility.

If a belief is unsupported, we should not force ourselves to accept it just because it comforts us.

A person who values truth must sometimes say, "I do not know," or "I cannot honestly believe this."

There is dignity in that honesty.

A clear reply is:

"I cannot honestly believe something just because it is comforting. Truth matters, even when it is difficult."

## A Calm Way to Continue the Conversation

Not every discussion about religion needs to become a debate.

Often, the best response is simple and respectful:

"I understand why this belief matters to you."

"I am not attacking you. I am examining the claim."

"I respect sincere believers, but I need evidence before accepting a belief."

"I do not think comfort and truth are the same."

"I would rather say I do not know than claim certainty without evidence."

These replies keep the conversation human.

The goal is not to defeat a person.

The goal is to stay honest.

## Appendix Summary

1. Many common objections to atheism rely on fear, tradition, comfort, or misunderstanding.
2. "You cannot disprove God" does not shift the burden of proof away from the believer.
3. Scientific uncertainty does not prove God.
4. Morality, meaning, beauty, and hope remain possible without religion.
5. Prayer, miracles, scripture, and personal experience do not provide reliable evidence for God.
6. Atheism is not another religion or another faith. It is disbelief due to insufficient evidence.
7. Honest doubt is more respectable than forced belief.

## Key Takeaway

Common objections to atheism often sound powerful because they are emotionally familiar. But when examined carefully, they do not overcome the basic problem: the personal God of religion lacks strong evidence. A calm, honest reply does not need anger. It needs clarity. Comfort is not proof, mystery is not evidence, and fear is not truth.

## Sources and Further Reading

1. Bertrand Russell, **Why I Am Not a Christian**.
2. Carl Sagan, **The Demon-Haunted World: Science as a Candle in the Dark**.

3. David Hume, **Dialogues Concerning Natural Religion**.
4. J. L. Mackie, **The Miracle of Theism**.
5. A. C. Grayling, **The God Argument**.
6. Michael Shermer, **The Believing Brain**.
7. Richard Dawkins, **The God Delusion**.
8. Daniel C. Dennett, **Breaking the Spell: Religion as a Natural Phenomenon**.

## Transition to Appendix C

The major objections have now been answered. The final appendix offers a short reading guide for readers who want to continue exploring science, atheism, morality, religion, and human meaning in greater depth.

# Appendix C: Recommended Further Reading

## Why Further Reading Matters

This book has presented the scientific case for atheism in simple, direct language. But no single book can cover every question fully.

A thoughtful reader may want to go deeper.

How did science replace supernatural explanation?
What is the strongest philosophical case against God?
How does evolution explain design?
What does neuroscience show about the mind?
How can morality stand without religion?
Why do religions arise in nearly every culture?
How can a person live meaningfully without belief in God?

The books listed here are not sacred authorities. They are not meant to be believed blindly. They are starting points for further thinking.

Read them as a free mind should read: carefully, critically, honestly, and with the willingness to change your view when evidence requires it.

## For Scientific Skepticism and Clear Thinking

### Carl Sagan, The Demon-Haunted World: Science as a Candle in the Dark

This is one of the best books for understanding why evidence matters. Sagan explains how human beings can be misled by superstition, fear, wishful thinking, and false certainty. He also shows why science is not cold or arrogant, but one of humanity's best tools for protecting truth.

### Michael Shermer, The Believing Brain

Shermer explains why people form beliefs first and often search for reasons afterward. This book is useful for understanding pattern-seeking, confirmation bias, and why religious belief can feel certain even when evidence is weak.

### James Randi, The Faith Healers

This book examines miracle claims, faith healing, and the danger of confusing hope with evidence. It is especially useful for readers interested in prayer, healing claims, and religious deception.

## For the Universe and Natural Reality

### Carl Sagan, Cosmos

A beautiful introduction to the scientific view of the universe. Sagan combines astronomy, history, science, and wonder in a way that helps readers feel the grandeur of a natural cosmos.

### Carl Sagan, Pale Blue Dot

This book places human life in cosmic perspective. It is especially powerful for readers who want to understand how a natural universe can inspire humility, responsibility, and awe without religious belief.

### Sean Carroll, The Big Picture

Carroll presents a modern scientific and philosophical view of reality. He explains how meaning, morality, consciousness, and purpose can be understood in a natural universe.

### Stephen Hawking, A Brief History of Time

A classic introduction to cosmology, time, black holes, and the universe. Some parts may be challenging, but the book remains important for readers interested in the universe without a simple creation story.

## For Evolution and Life Without Design

### Charles Darwin, On the Origin of Species

This is the foundational work of evolutionary biology. The language is older, but the central insight remains powerful: natural selection can explain the appearance of design without a designer.

**Richard Dawkins, The Blind Watchmaker**

A clear and forceful explanation of how evolution can produce complex life without divine design. This book directly addresses the common claim that life is too complex to evolve naturally.

**Jerry A. Coyne, Why Evolution Is True**

An excellent modern introduction to the evidence for evolution. It covers fossils, genetics, anatomy, biogeography, and natural selection in an accessible way.

**Neil Shubin, Your Inner Fish**

Shubin shows how the human body carries the history of earlier life forms. This book is especially helpful for understanding common ancestry and why our bodies contain evolutionary traces.

**Sean B. Carroll, Endless Forms Most Beautiful**

A fascinating book on evolutionary development. It explains how changes in genetic regulation helped produce the diversity of animal forms.

## For the Body, Disease, and Imperfect Design

**Randolph M. Nesse and George C. Williams, Why We Get Sick**

This book explains disease from an evolutionary perspective. It is especially useful for understanding why the human body is not perfectly designed and why vulnerability is built into biology.

**Randolph M. Nesse, Good Reasons for Bad Feelings**

Nesse applies evolutionary thinking to emotions and mental suffering. This book is useful for readers interested in anxiety, depression, fear, and human vulnerability.

**Siddhartha Mukherjee, The Emperor of All Maladies**

A deeply human and scientific history of cancer. It helps readers understand cancer as biology, not divine punishment.

**Siddhartha Mukherjee, The Gene**

A broad and readable history of genetics. It is helpful for understanding heredity, mutation, genetic disease, and the biological basis of life.

## For Mind, Brain, and the Soul Question

**Antonio Damasio, Descartes' Error**

Damasio shows how emotion, reason, body, and brain are deeply connected. This book is useful for readers questioning the idea of a mind separate from the body.

**Antonio Damasio, Self Comes to Mind**

A deeper exploration of consciousness and the self. It helps explain how the sense of “I” can arise from brain and body processes.

**Oliver Sacks, The Man Who Mistook His Wife for a Hat**

A humane and memorable collection of neurological cases. Sacks shows how changes in the brain can alter perception, memory, identity, and personality.

**Eric R. Kandel, In Search of Memory**

A major work on memory, brain science, and the biology of mind. It is especially useful for understanding why memory is not a spiritual substance, but a biological process.

**Stanislas Dehaene, Consciousness and the Brain**

A modern scientific account of consciousness. It is useful for readers who want to understand how brain activity relates to awareness.

**Patricia S. Churchland, Touching a Nerve**

A clear and accessible discussion of the self as brain-based. Churchland also addresses morality, responsibility, and human nature from a natural perspective.

## For the Philosophy of Atheism

### Bertrand Russell, Why I Am Not a Christian

A short and classic work. Russell examines arguments for God, morality, fear, and religious authority with clarity and courage.

### David Hume, Dialogues Concerning Natural Religion

A major philosophical work on the design argument, suffering, and the limits of religious reasoning. Some readers may find the style older, but the ideas remain important.

### J. L. Mackie, The Miracle of Theism

A strong philosophical examination of arguments for and against God. It is more demanding than some other books, but very valuable for serious readers.

### A. C. Grayling, The God Argument

A clear modern discussion of atheism, humanism, reason, and morality. It is accessible and helpful for readers who want a broad overview.

### Julian Baggini, Atheism: A Very Short Introduction

A concise introduction to atheism. It is useful for readers who want a short, balanced starting point.

## For Suffering, Evil, and Divine Hiddenness

**Bart D. Ehrman, God's Problem**

Ehrman examines the problem of suffering through the Bible and religious belief. It is helpful for readers struggling with the moral difficulty of innocent pain.

**William L. Rowe, "The Problem of Evil and Some Varieties of Atheism"**

This essay is one of the most important modern discussions of the problem of evil. It is more philosophical, but central to the debate.

**J. L. Schellenberg, Divine Hiddenness and Human Reason**

Schellenberg's work is essential for understanding the argument from divine hiddenness. It asks why a loving God would remain unclear to sincere seekers.

**Theodore M. Drange, Nonbelief and Evil**

This book examines both suffering and nonbelief as challenges to theism. It is useful for readers who want a more detailed philosophical approach.

## For Religion as a Human Phenomenon

**Daniel C. Dennett, Breaking the Spell: Religion as a Natural Phenomenon**

Dennett examines religion as something that can be studied scientifically and culturally. The book is useful for understanding religion without assuming it comes from God.

**Pascal Boyer, Religion Explained**

Boyer explores how human minds naturally form religious ideas. This book is helpful for readers interested in psychology, culture, and the origin of religious belief.

**Émile Durkheim, The Elementary Forms of Religious Life**

A classic sociological study of religion. It is older and more academic, but important for understanding religion as a social force.

**David Sloan Wilson, Darwin's Cathedral**

Wilson examines religion through evolutionary and group-level social functions. Readers may not agree with every point, but the book is useful for understanding how religion can bind communities.

**James George Frazer, The Golden Bough**

A classic study of myth, ritual, and religion. Some parts are outdated, but it remains historically important.

## For Morality and Humanism Without God

**David Hume, An Enquiry Concerning the Principles of Morals**

Hume argues that morality is deeply connected to human feeling, sympathy, and social life. This is an important work for understanding morality without divine command.

**John Stuart Mill, Utilitarianism**

Mill presents a moral theory based on well-being and consequences. Readers may not agree with all of it, but it remains central to secular moral thought.

**Peter Singer, The Expanding Circle**

Singer discusses how moral concern can widen beyond tribe, nation, religion, and species. This book is useful for understanding moral progress without God.

**Frans de Waal, Primates and Philosophers**

De Waal explores the roots of morality in social animals. This book helps show that cooperation, empathy, and fairness have natural roots.

**Paul Kurtz, The Courage to Become**

A humanist work on meaning, responsibility, and living with courage without supernatural belief.

## For Meaning, Wonder, and Life Without God

**Richard Dawkins, Unweaving the Rainbow**

This book answers the fear that science destroys beauty. Dawkins argues that scientific understanding can deepen wonder rather than reduce it.

**Carl Sagan, Pale Blue Dot**

Worth mentioning again because it is one of the most powerful books for understanding human responsibility in a vast natural universe.

**Bertrand Russell, The Conquest of Happiness**

A practical and thoughtful book on living well. It is not only about atheism, but it reflects a secular approach to human flourishing.

**Sam Harris, Waking Up**

A secular discussion of meditation, consciousness, and spirituality without religion. Some readers may disagree with parts of it, but it is useful for thinking about inner life without supernatural belief.

**A. C. Grayling, The Meaning of Things**

A collection of short reflections on life, ethics, loss, love, and human meaning from a secular perspective.

# A Suggested Reading Path

Readers do not need to read all these books.

A simple path could begin like this:

First, read **The Demon-Haunted World** by Carl Sagan to strengthen clear thinking.

Second, read **Why Evolution Is True** by Jerry Coyne to understand life without design.

Third, read **Why We Get Sick** by Nesse and Williams to understand the body as a product of evolution.

Fourth, read **The Man Who Mistook His Wife for a Hat** by Oliver Sacks to feel the connection between brain and self.

Fifth, read **Why I Am Not a Christian** by Bertrand Russell for a classic statement of unbelief.

Sixth, read **The God Argument** by A. C. Grayling for a modern overview of atheism and humanism.

Seventh, read **Pale Blue Dot** by Carl Sagan to bring science, humility, and human responsibility together.

This path gives a strong foundation without becoming too technical.

## How to Read Critically

Further reading should not become a new form of obedience.

Do not believe an author simply because he or she agrees with atheism.

Ask questions.

What evidence is being offered?
Is the argument fair?
Are opposing views represented honestly?
Is the author making claims stronger than the evidence allows?
Does the explanation fit reality better than alternatives?
Could future evidence change the conclusion?

A free mind does not replace one authority with another.

It learns how to think.

That is the real purpose of reading.

## A Final Note to the Reader

The end of this book is not the end of inquiry.

It may be the beginning of a clearer intellectual life.

To question God honestly is not a small step. For some readers, it may touch family, identity, fear, grief, memory, and hope. Move carefully. Think patiently. Read widely. Be honest with yourself. Be kind to others.

Truth does not need panic.

Reality does not need decoration to be profound.

A natural universe is still astonishing.

A finite life is still precious.

A human mind is still beautiful.

Compassion is still urgent.

The work of understanding continues.

# Appendix Summary

1. Further reading helps readers explore science, atheism, morality, religion, mind, suffering, and meaning in greater depth.
2. No author should be treated as unquestionable authority.
3. Scientific and philosophical reading should strengthen clarity, not create a new dogma.
4. The best reading path begins with clear thinking, then moves through evolution, disease, mind, atheism, humanism, and cosmic perspective.
5. A free mind reads critically, honestly, and with courage.

# Key Takeaway

The search for truth does not end with this book. It continues through evidence, reason, reading, reflection, and honest conversation. Science-based atheism is not a closed doctrine. It is a commitment to follow reality wherever it leads, without fear and without the need for God.

# Thank You and Please Review the Book

Thank you for taking the time to read this book. I sincerely hope it has encouraged you to think more deeply, question more honestly, and look at religion, science, and human belief with greater clarity.

My goal in writing the **No God Required Series** is simple: to present serious, science-based, and easy-to-understand ideas about God, religion, nature, morality, and human meaning. These books are written for readers who value evidence, reason, intellectual honesty, and the courage to ask difficult questions.

If you found this book thoughtful, helpful, or worth reading, I would be truly grateful if you could leave a brief and honest review. Even a few sincere lines can make a meaningful difference. Your review helps other readers discover this series and encourages more open, rational, and evidence-based discussion about some of life's deepest questions.

Thank you again for your time, your trust, and your support.

Warm regards,
**Dr. Prabhat Das**

# OTHER BOOKS IN THIS SERIES

**No God Required: How Science Explains Reality Without the Divine** is a bold, clear, science-based series by **Dr. Prabhat Das**. It reveals why humans created gods, how science made supernatural explanations unnecessary, and how meaning, morality, courage, and wonder can flourish without the divine.

**The 7 Books in This Series: -**

Book 1: **Why and How Humans Created God**

Book 2: **How Science Made God Unnecessary**

Book 3: **The Universe Without a Creator**

Book 4: **Life Without a Designer**

Book 5: **Mind Without Soul, Morality Without God**

Book 6: **The Failure of the Personal God**

Book 7: **The Scientific Case for Atheism**

www.ingramcontent.com/pod-product-compliance
Lightning Source LLC
LaVergne TN
LVHW011702200926
848484LV00004B/1038
* 9 7 9 8 9 5 1 0 6 7 1 8 0 *